CLINICAL NUTRITION AND AGING

Sarcopenia and Muscle Metabolism

CLINICAL NUTRITION AND AGING

Sarcopenia and Muscle Metabolism

Edited by
Chad Cox, PhD

APPLE ACADEMIC PRESS

Apple Academic Press Inc. | Apple Academic Press Inc.
3333 Mistwell Crescent | 9 Spinnaker Way
Oakville, ON L6L 0A2 | Waretown, NJ 08758
Canada | USA

©2016 by Apple Academic Press, Inc.

First issued in paperback 2021

Exclusive worldwide distribution by CRC Press, a member of Taylor & Francis Group

No claim to original U.S. Government works

ISBN 13: 978-1-77463-619-0 (pbk)
ISBN 13: 978-1-77188-370-2 (hbk)

Library and Archives Canada Cataloguing in Publication

Clinical nutrition and aging : sarcopenia and muscle metabolism / edited by Chad Cox, PhD.

Issued in print and electronic formats.
ISBN 978-1-77188-370-2 (hardcover).--ISBN 978-1-77188-371-9 (pdf)
1. Muscular atrophy--Age factors. 2. Muscular atrophy--Diet therapy. 3. Muscular atrophy--Prevention. 4. Muscular atrophy--Treatment. 5. Muscles--Metabolism. 6. Aging--Nutritional aspects. 7. Older people--Nutrition. I. Cox, Chad (Chad L.), editor

RC935.A8C65 2016   616.7'4   C2015-906725-1   C2015-906726-X

Library of Congress Cataloging-in-Publication Data

Names: Cox, Chad (Chad L.), editor.
Title: Clinical nutrition and aging : sarcopenia and muscle metabolism / Chad Cox.
Description: Toronto; New Jersey : Apple Academic Press, 2015. | Includes
Includes bibliographical references and index.
Identifiers: LCCN 2015037195 | ISBN 9781771883702 (alk. paper)
Subjects: LCSH: Nutrition disorders in old age. | Older people--Diseases--Nutritional aspects. | Older people--Nutrition.
Classification: LCC RC620.6 .C69 2015 | DDC 618.97/639--dc23
LC record available at http://lccn.loc.gov/2015037195

Apple Academic Press also publishes its books in a variety of electronic formats. Some content that appears in print may not be available in electronic format. For information about Apple Academic Press products, visit our website at **www.appleacademicpress.com** and the CRC Press website at **www.crcpress.com**

About the Editor

CHAD COX, PhD

Dr. Chad L. Cox is a Lecturer in the Department of Chemistry and the Department of Family and Consumer Sciences at California State University, Sacramento. He also teaches at Sacramento City College and the University of California, Davis. He holds a PhD in Nutritional Biology, a Bachelor of Science in Exercise Biology, and a Bachelor of Science in Nutrition Science, all from UC Davis. His research interests include the causes of obesity and obesity-related chronic diseases, how exercise training can induce changes in the regulation of gene expression that can lead to improvements in insulin sensitivity and promote energy balance, and the development of pharmacological agents that could help reduce the epidemic of obesity, Type 2 diabetes, and metabolic syndrome.

Contents

Acknowledgment and How to Cite

The editor and publisher thank each of the authors who contributed to this book. The chapters in this book were previously published elsewhere. To cite the work contained in this book and to view the individual permissions, please refer to the citation at the beginning of each chapter. Each chapter was carefully selected by the editor; the result is a book that looks at nutrition and aging from a variety of perspectives. The chapters included are broken into three sections, which describe the following topics:

- Chapter 1: Early nutritional intervention is more effective at preventing or delaying sarcopenia than waiting until old age to improve composition of the diet.
- Chapter 2: When muscle mass and nutrition are assessed concurrently, instead of as separate issues, this allows for a more comprehensive understanding of sarcopenia and makes it easier to select appropriate interventions.
- Chapter 3: This article provides a comprehensive summary of the current research regarding the correct intake of proteins and other nutrients, such as antioxidants, vitamin D, and fatty acids in elderly populations.
- Chapter 4: Increased amino acid consumption contributes to muscle protein synthesis and reduction of adipose mass in calorically restricted elderly adults.
- Chapter 5: Increased consumption of Branched Chain Amino Acids may help overcome anabolic resistance, and have a favorable effect on muscle protein synthesis and subsequent maintenance of muscle mass.
- Chapter 6: Consumption of Whey protein stimulates greater rates of muscle protein synthesis than does consumption of soy protein, presumably due to differences in leucine content and digestibility.
- Chapter 7: Older adults need greater amounts of protein than what is currently recommended. This article provides evidence in favor of raising the current RDA for protein in older adults.
- Chapter 8: Exercise enhances mitochondrial function, biogenesis, dynamics, turnover, and quality control in aging muscle tissue, which contributes to the efficacy of exercise as a preventive and treatment intervention for sarcopenia.

- Chapter 9: Due to advances in our understanding of muscle biology and continued increase, more effective research investigating the pharmacological treatment of sarcopenia is now feasible and. Currently, ursolic acid and EPA supplements show much promise as effective therapeutic strategies.
- Chapter 10: This article discusses the concept of decreased "anabolic threshold" in relation to muscle wasting and explains how the time of feeding and quantity/quality of the protein fed could be important for normalizing protein metabolism in individuals with sarcopenia.
- Chapter 11: Despite the potential efficacy of antioxidant supplementation in the treatment of sarcopenia, we need further studies to improve our understanding of oxidative mechanisms and the pharmacokinetics of antioxidant supplementation, along with appropriate methodological approaches, before we can justify the widespread use of antioxidant supplements.
- Chapter 12: Hormone replacement therapy (combined with nutrition and exercise therapy) may be an effective treatment for sarcopenia; however more clinical data is needed before effective dosages can be determined.

List of Contributors

Deborah Agostini
Department of Biomolecular Sciences, Division of Exercise and Health Sciences, University of Urbino Carlo Bo, Via A. Saffi 2, 61029 Urbino, Italy

Francesca Allieri
Department of Public Health, Experimental and Forensic Medicine, School of Medicine, University of Pavia and Endocrinology and Nutrition Unit, Azienda di Servizi alla Persona di Pavia, Via Emilia 12, 27100 Pavia, Italy

Carolyn J. Alish
Scientific and Medical Affairs, Abbott Nutrition, Abbott Laboratories, 3300 Stelzer Road, Columbus, OH 43219, USA

Giosuè Annibalini
Department of Biomolecular Sciences, Division of Exercise and Health Sciences, University of Urbino Carlo Bo, Via A. Saffi 2, 61029 Urbino, Italy

Carla Basualto-Alarcón
Programa de Anatomía y Biología del Desarrollo, Facultad de Medicina, Instituto de Ciencias Biomédicas, Universidad de Chile, Santiago, Chile

Elena Barbieri
Department of Biomolecular Sciences, Division of Exercise and Health Sciences, University of Urbino Carlo Bo, Via A. Saffi 2, 61029 Urbino, Italy

Jamie I. Baum
Department of Food Science, University of Arkansas, 2650 N. Young Ave., Fayetteville, AR 72704, USA

Leigh Breen
Exercise Metabolism Research Group, Department of Kinesiology, McMaster University, 1280 Main St. West, Hamilton, ON, Canada, L8S 4K1

Nicholas A. Burd
Exercise Metabolism Research Group, Department of Kinesiology, McMaster University, 1280 Main St. West, Hamilton, ON, Canada, L8S 4K1

Darren G. Candow
Faculty of Kinesiology and Health Studies, University of Regina, Regina, Saskatchewan, S4S 0A2, Canada

Francesco Cerullo
Dipartimento di Scienze Gerontologiche, Geriatriche e Fisiatriche, Università Cattolica del Sacro Cuore, Largo Francesco Vito 1, 00168 Roma, Italy

Matteo Cesari
Institut du Vieillissement, Université de Toulouse, 31000 Toulouse, France

Tyler A. Churchward-Venne
Exercise Metabolism Research Group, Department of Kinesiology, McMaster University, 1280 Main St. West, Hamilton, ON, Canada, L8S 4K1

Robert H. Coker
Center for Translational Research in Aging and Longevity, Little Rock, AR, USA

Cyrus Cooper
MRC Lifecourse Epidemiology Unit (University of Southampton), Southampton General Hospital, Southampton SO16 6YD, UK

Dominique Dardevet
Clermont Université and Unité de Nutrition Humaine, Université d'Auvergne, BP 10448, 63000 Clermont-Ferrand, France, INRA, UMR 1019, UNH, CRNH Auvergne, 63000 Clermont-Ferrand, France, and UNH Centre de Clermont-Ferrand-Theix, INRA, 63122 Saint Genès, France

Mauro De Santi
Department of Biomolecular Sciences, Division of Exercise and Health Sciences, University of Urbino Carlo Bo, Via A. Saffi 2, 61029 Urbino, Italy

Nicolaas Deutz
Center for Translational Research in Aging and Longevity, Little Rock, AR, USA

Javier Duran
Programa de Fisiología y Biofísica, Facultad de Medicina, Instituto de Ciencias Biomédicas, Universidad de Chile, Santiago, Chile

Manuel Estrada
Programa de Fisiología y Biofísica, Facultad de Medicina, Instituto de Ciencias Biomédicas, Universidad de Chile, Santiago, Chile

Milena Faliva
Department of Public Health, Experimental and Forensic Medicine, School of Medicine, University of Pavia and Endocrinology and Nutrition Unit, Azienda di Servizi alla Persona di Pavia, Via Emilia 12, 27100 Pavia, Italy

Scott C. Forbes
Faculty of Physical Education & Recreation, University of Alberta, Edmonton, Alberta, T6G 2R3, Canada

Giovanni Gambassi
Dipartimento di Scienze Gerontologiche, Geriatriche e Fisiatriche, Università Cattolica del Sacro Cuore, Largo Francesco Vito 1, 00168 Roma, Italy

Michele Guescini
Department of Biomolecular Sciences, Division of Exercise and Health Sciences, University of Urbino Carlo Bo, Via A. Saffi 2, 61029 Urbino, Italy

Refaat A. Hegazi
Scientific and Medical Affairs, Abbott Nutrition, Abbott Laboratories, 3300 Stelzer Road, Columbus, OH 43219, USA

Jonathan P. Little
School of Health and Exercise Sciences, University of British Columbia Okanagan, Kelowna, British Columbia, V1V 1V7, Canada

Francesco Lucertini
Department of Biomolecular Sciences, Division of Exercise and Health Sciences, University of Urbino Carlo Bo, Via A. Saffi 2, 61029 Urbino, Italy

Rodrigo Maass
Facultad de Medicina, Departamento de Morfofunción, Universidad Diego Portales, Santiago, Chile

Ralph J. Manders
Exercise Physiology Research Group, Department of Kinesiology, KU Leuven, Heverlee, B-3001, Belgium

Sharon Miller
Department of Geriatrics, University of Arkansas for Medical Sciences, and Healthspan, LLC, Little Rock, AR, USA

Francesca Monteferrario
Department of Public Health, Experimental and Forensic Medicine, School of Medicine, University of Pavia and Endocrinology and Nutrition Unit, Azienda di Servizi alla Persona di Pavia, Via Emilia 12, 27100 Pavia, Italy

Laurent Mosoni
Clermont Université and Unité de Nutrition Humaine, Université d'Auvergne, BP 10448, 63000 Clermont-Ferrand, France and INRA, UMR 1019, UNH, CRNH Auvergne, 63000 Clermont-Ferrand, France

Isabelle Papet
Clermont Université and Unité de Nutrition Humaine, Université d'Auvergne, BP 10448, 63000 Clermont-Ferrand, France and INRA, UMR 1019, UNH, CRNH Auvergne, 63000 Clermont-Ferrand, France

Simone Perna
Department of Public Health, Experimental and Forensic Medicine, School of Medicine, University of Pavia and Endocrinology and Nutrition Unit, Azienda di Servizi alla Persona di Pavia, Via Emilia 12, 27100 Pavia, Italy

Gabriella Peroni
Department of Public Health, Experimental and Forensic Medicine, School of Medicine, University of Pavia and Endocrinology and Nutrition Unit, Azienda di Servizi alla Persona di Pavia, Via Emilia 12, 27100 Pavia, Italy

Marie-Agnès Peyron
Clermont Université and Unité de Nutrition Humaine, Université d'Auvergne, BP 10448, 63000 Clermont-Ferrand, France and INRA, UMR 1019, UNH, CRNH Auvergne, 63000 Clermont-Ferrand, France

Stuart M. Phillips
Exercise Metabolism Research Group, Department of Kinesiology, McMaster University, 1280 Main St. West, Hamilton, ON, Canada, L8S 4K1

Emanuela Polidori
Department of Biomolecular Sciences, Division of Exercise and Health Sciences, University of Urbino Carlo Bo, Via A. Saffi 2, 61029 Urbino, Italy

Lucia Potenza
Department of Biomolecular Sciences, Division of Exercise and Health Sciences, University of Urbino Carlo Bo, Via A. Saffi 2, 61029 Urbino, Italy

Didier Rémond
Clermont Université and Unité de Nutrition Humaine, Université d'Auvergne, BP 10448, 63000 Clermont-Ferrand, France and INRA, UMR 1019, UNH, CRNH Auvergne, 63000 Clermont-Ferrand, France

Erica Repaci
Department of Public Health, Experimental and Forensic Medicine, School of Medicine, University of Pavia and Endocrinology and Nutrition Unit, Azienda di Servizi alla Persona di Pavia, Via Emilia 12, 27100 Pavia, Italy

Siân Robinson
MRC Lifecourse Epidemiology Unit (University of Southampton), Southampton General Hospital, Southampton SO16 6YD, UK

Mariangela Rondanelli
Department of Public Health, Experimental and Forensic Medicine, School of Medicine, University of Pavia and Endocrinology and Nutrition Unit, Azienda di Servizi alla Persona di Pavia, Via Emilia 12, 27100 Pavia, Italy

Kunihiro Sakuma
Research Center for Physical Fitness, Sports and Health, Toyohashi University of Technology, 1-1 Hibarigaoka, Tenpaku-cho, Toyohashi 441-8580, Japan

Abby C. Sauer
Scientific and Medical Affairs, Abbott Nutrition, Abbott Laboratories, 3300 Stelzer Road, Columbus, OH 43219, USA

Isabelle Savary-Auzeloux
Clermont Université and Unité de Nutrition Humaine, Université d'Auvergne, BP 10448, 63000 Clermont-Ferrand, France and INRA, UMR 1019, UNH, CRNH Auvergne, 63000 Clermont-Ferrand, France

Avan Aihie Sayer
MRC Lifecourse Epidemiology Unit (University of Southampton), Southampton General Hospital, Southampton SO16 6YD, UK

Scott Schutzler
Center for Translational Research in Aging and Longevity, Little Rock, AR, USA

Laura Stocchi
Department of Biomedicine and Prevention, University of Tor Vergata, Via Montpellier 1, 00133 Rome, Italy

Vilberto Stocchi
Department of Biomolecular Sciences, Division of Exercise and Health Sciences, University of Urbino Carlo Bo, Via A. Saffi 2, 61029 Urbino, Italy

Mark A. Tarnopolsky
Michael G. DeGroote School of Medicine, Pediatrics and Neurology, McMaster University, Hamilton, ON, Canada

Maurits F. J. Vandewoude
Department of Geriatrics, ZNA St. Elisabeth Leopoldstraat 26, University of Antwerp, 2000 Antwerp, Belgium

Diego Varela
Programa de Fisiología y Biofísica, Facultad de Medicina, Instituto de Ciencias Biomédicas, Universidad de Chile, Santiago, Chile

Robert R. Wolfe
Department of Geriatrics, University of Arkansas for Medical Sciences, 4301 W. Markham St. #806, Little Rock, AR 72705, USA

Akihiko Yamaguchi
School of Dentistry, Health Sciences University of Hokkaido, Kanazawa, Ishikari-Tobetsu, Hokkaido 061-0293, Japan

Yifan Yang
Exercise Metabolism Research Group, Department of Kinesiology, McMaster University, 1280 Main St. West, Hamilton, ON, Canada, L8S 4K1

Introduction

Sarcopenia is the loss of muscle mass and strength that occurs as people age. It is difficult to know exactly how many older people experience sarcopenia, since different clinicians define it differently. According to some estimates (Sakuma and Yamaguchi, 2012), the prevalence of sarcopenia ranges from 13 percent to 24 percent in adults over sixty years of age to more than 50 percent in people eighty and older; other research (Cruz-Jentoft, 2014) indicates the international prevalence of sarcopenia is between 1 and 29 percent in community-dwelling populations, and between 14 and 33 percent in long-term care populations.

Regardless of the exact numbers, sarcopenia is a very real and common problem among older adults. It affects mobility, energy levels, and other aspects of health and well-being, and it contributes to decreased survival rates after a critical illness. In the article by Sakuma and Yamaguchi, they report that the estimated direct health-care costs attributable to sarcopenia in the United States in 2000 were $18.5 billion ($10.8 billion in men and $7.7 billion in women), which represented about 1.5 percent of total health-care expenditures for that year. For economic reasons alone, this is a serious problem.

As individuals and clinicians, it is an equally serious issue. We can expect that at seventy we will not be as strong as we were at thirty. We assume that at ninety, we will be even weaker than we were at seventy. This gradual loss of muscle is perceived by most of us as "normal." To an extent this is true—and yet not everybody gets as weak and frail. Some people retain a certain hardiness even into their eighties and nineties. Why is this so?

The implication is that there must be factors that influence the varying degrees with which different individuals experience sarcopenia. Some of these factors may be beyond our control—"good genes," for example—but others may have to do with diet and lifestyle, factors that are in our control (at least more so than our genetic material).

Regrettably, poor diet is one of the most common problems practitioners encounter when treating older adults. Many individuals in this population have low nutrient intakes, for a variety of reasons that range from physical deficits to economic hardship. Dental problems in the elderly may make them more likely to choose softer foods that often lack protein; delayed gastric emptying can reduce appetite; hormonal changes may cause longer-lasting feelings of satiety. On top of that, meat is more expensive than foods rich in starch, and for this reason individuals who are living on a fixed income may tend to fill up on cheap, processed carbohydrates. Lack of physical strength may also make packaged, processed foods more appealing.

But these issues related to consuming poor diet are not impossible to overcome. If we can understand the lifestyle factors that influence the rate of decline in muscle mass and strength in older age, we can develop practical strategies that will help to prevent or delay sarcopenia, allowing people to maintain a higher quality of life into old age. In the first paper in this compendium, Robinson et al. consider the evidence that links diet to muscle mass and strength in the elderly. They discuss the evidence that supports the potential importance of diets of adequate quantity and quality, which the authors break down into sufficient intakes of protein, vitamin D, and antioxidant nutrients. Although much of this evidence is observational—and they are unable to describe the specific mechanisms that might help prevent or delay sarcopenia—the authors do offer us a valuable key point: sarcopenia prevention needs to begin before old age. Research confirms that the greater the peak strength attained during a person's younger adult life, the more likely an individual will have greater strength in their older years. Early intervention can make a difference. We need to teach our students and patients that if they optimize their nutrition now, they will be investing in their future well-being.

In the second article, Vandewoude et al. look at this issue by defining two categories of sarcopenia: Primary sarcopenia, when no specific cause can be identified, is progressive and associated with the impact of aging. Changes are seen at the cellular level, with changes in motor neurons and mitochondrial function, as well at the hormonal level (resulting in insulin resistance, an increase in proinflammatory cytokines, etc.). Secondary sarcopenia is the result of both lifestyle (lack of physical activity, poor diet) and secondary physical effects (chronic inflammation, for example). Sarcopenia, then, is both an outcome and a process. As an outcome, it is a diagnosis that causes frailty and mobility issues in older adults. As an active process, it is going on inside the body of every adult.

Instead of focusing on good nutrition, these authors focus on malnutrition, which they initially define as "an imbalance of energy, protein, and other nutrients that cause measurable negative effects on body composition, physical function, and clinical outcomes." They then further develop this definition by using new guidelines from the International Guideline Consensus Committee, which link diet with physical results using three categories:

1. starvation-related without inflammation
2. chronic disease or conditions that impose sustained mild-to-moderate inflammation
3. acute disease or injury states, when inflammatory response is marked

Sarcopenia would fall under category number 2.

Looking at sarcopenia from this perspective allows us to create better assessment methods. Measuring lean body mass is a useful assessment of sarcopenia, since skeletal muscle mass, which provides strength, mobility, and balance, is the primary constituent of lean body mass. Looking at muscle mass also helps us better understand secondary sarcopenia, since muscle mass plays a critical role in whole-body protein metabolism. Historically, patients have been screened for either malnutrition or sarcopenia, but rarely for both conditions at the same time. Vandewoude and his colleagues call for more comprehensive screening methodology that will aid more targeted interventions.

In the third article, Rondanelli and her colleagues focus more deeply on secondary sarcopenia. They discuss how muscle weakness makes it difficult for elderly people to walk—which makes them less likely to walk, which in turn impacts their cardiovascular system. At the same time, if they do walk, they are more likely to fall, putting them at risk for bone fractures. The authors reinforce the biochemical effects of sarcopenia, and discuss this in more detail than was offered in the previous article.

Sarcopenia alters protein metabolism; protein synthesis is no longer adequate. This in turn means that muscle cells progressively lose their sensitivity to anabolic stimuli, which is what causes hormonal issues. Ultimately, the authors of this article give us a good summary of the current research on elderly populations regarding the correct intake of proteins and other nutrients, such as antioxidants, vitamin D, and fatty acids. They discuss the role of dietary supplements in the prevention and treatment of sarcopenia.

In the next section of the compendium, we turn our attention to protein supplementation and exercise as sarcopenia interventions. In the book's fourth article, Coker and his colleagues add another important element to our discussion: obesity in the elderly. Lack of activity brought on by muscle weakness contributes to obesity—and obesity makes it more difficult for older individuals to move around. And it's not easily stopped, since "going on a diet"—in the sense of restricting calories—may only accelerate sarcopenia's progress. Based on these observations, combined with the results of their research, the authors call for increased amino acid consumption in elderly populations to contribute to and increase stimulation of muscle protein synthesis. They also recommend protein supplements that are high-potency formulations of whey protein and amino acids.

The dysfunction of skeletal muscle protein synthesis and hormonal disturbances that have already been mentioned are sometimes referred to as anabolic resistance. In the fifth article, Manders et al. discuss data suggesting that that high doses of leucine may help overcome anabolic resistance, and have a favorable effect on muscle protein synthesis and subsequent maintenance of muscle mass. Leucine contributes to amino acid availability for muscle protein synthesis, inhibits muscle protein breakdown resulting in a gradually greater net muscle protein balance, and also enhances glucose disposal to help maintain blood glucose homeostasis.

Both amino acid supplementation and resistance exercise are examples of proven effective sarcopenia interventions. In the sixth article, Yifan Yang et al. explain that the source of proteins is also an important factor. They report that whey protein (already mentioned by Coker et al.) and cow's milk stimulate greater rates of muscle protein synthesis than do proteins from sources such as micellar casein or soy—and they base this finding on assessments of both resting muscles and following resistance exercise. They conclude that the digestion kinetics and amino acid content of these proteins is markedly different. If supported by further research, this is an important piece of information for anyone who provides nutritional counseling to the elderly.

In the next article, Baum and Wolfe point out that older adults may need greater amounts of protein than currently being recommended. As earlier papers indicated, higher protein intakes are associated with increased muscle protein synthesis, which is correlated with increased muscle mass and function. This, in turn, is linked to improved physical function and would presumably also be associated with a delay in the onset and progression of sarcopenia.

In the eighth article, Barbieri and her colleagues give us some more valuable information. They point out that recent research indicates that adults over sixty spend as many as eight to twelve hours per day in sedentary pursuits. This inactivity reduces aerobic capacity, while it accelerates muscle catabolism, mitochondrial dysfunction, and oxidative stress. The authors state that these problems can lead to a "vicious circle" of muscle loss, injury, and inefficient repair mechanisms that contribute to elderly people becoming even more sedentary. Barbieri et al. call for regular exercise programs for elderly populations, which will likely reduce and/or prevent many of the factors that contribute to the vicious circle. They recommend a combination of endurance (aerobic) and strength training, pointing out that aerobic/endurance exercise helps to maintain and improve cardiovascular and respiratory function, while strength/resistance training improves muscle strength and function. Exercise programs like this would likely also contribute to reducing the prevalence of other chronic diseases in the elderly, such as cardiovascular disease, diabetes, and osteoporosis, leading to significant improvements in the quality of life of many senior citizens.

The authors conclude that exercise should be a "medical prescription" for most older adults.

In the third section of this compendium, we turn to other less traditional therapeutic strategies for the treatment of sarcopenia. The first article in this section by Sakuma and Yamaguchi further underlines the fact that since skeletal muscle has such an essential role in mobility and metabolic function, any deterioration in the contractile, material, and metabolic

properties of skeletal muscle would presumably have extremely significant effect on health and well-being. The authors point not only to resistance training, amino acid supplements, in particular leucine as potential
sarcopenia interventions but also to other more recently discovered pos-
sibilities including ursolic acid and EPA supplements. Due to advances in
muscle biology research, they point out that pharmacological treatment is
now a feasible and much-needed research direction.

In the next article, Dardavet and her colleagues delve more deeply into
how to treat sarcopenia with nutrition. Eating changes protein metabolism
during the day; whole-body proteins are stored after eating and then lost
in later periods. In a healthy adult, the gains and losses balance out, but
in older individuals, the loss is often not recovered. Therefore, the time of
feeding could be important for this population. At a more technical level,
they suggest two strategies to be used either alone or in combination:

1. increase the anabolic signals, in particular amino acid availability
2. increase the efficiency of the after-meal period with strategies aim-
    ing at decreasing the "anabolic threshold" of skeletal muscle.

In chapter 11, Cerullo and his colleagues discuss the "free radical
theory of aging"—in other words, that oxidative damage is one of the
major contributors to the decline in skeletal muscle mass that occurs as
people age. They remind us that if free radicals contribute to the aging
process, then inhibiting them might slow the aging process in terms of
maintaining muscle strength. If substances with antioxidant capacities
can counteract oxidative damage, they may play a key role in preventing the onset of age-related sarcopenia. Although there is some evidence
that oral antioxidant supplementation may reduce muscle damage, the
authors acknowledge that experimental results are largely preliminary
and clinically relevant. In fact, a large body of evidence seems to indicate that we should be very cautious about using antioxidant supplements on a large scale in the hopes of slowing the aging process. The
authors call for further studies to improve our understanding of oxidative
mechanisms, and the pharmacokinetics of antioxidant supplementation,
along with appropriate methodological approaches. Until this research
has been done, they caution against the widespread use of antioxidant
supplements.

Finally, in the last article, Basualto-Alarcón and his colleagues discuss the fact that our still incomplete understanding of sarcopenia means that we do not understand the variable outcomes of different treatment approaches, including exercise, nutrition, and hormones. The authors are interested in androgen supplementation as a treatment option (combined with exercise and nutritional therapy), since skeletal muscle is a target tissue for anabolic steroids, but they acknowledge that at this point the evidence linking testosterone deficiency and sarcopenia are mostly only observational rather than clinical. They call for studies that target the effects of anabolic steroids at the cellular level with in vivo models, in order to broaden our understanding of the role androgens play in skeletal muscle function in older people populations.

These twelve articles suggest treatment directions for practitioners working with older populations—but even more, these articles lay the foundation for future research which will further our understanding of sarcopenia and its treatment. This is a vital area of research because in most parts of the world, the aged population is growing rapidly. Basualto-Alarcón et al. point out that in the United States alone, longer life spans and the aging of baby boomers will double the over-sixty population over the next three decades. By 2030, twenty percent of the US population will be made up of adults. As this older population group loses muscle strength, they will also experience increased morbidity and mortality—either directly or indirectly via the development of secondary diseases such as diabetes, obesity, and cardiovascular disease.

Aging, however, does not have to equal poor health and weakness. The increased life expectancy in developed and developing countries should be interpreted as a call to identify interventions that can effectively slow or prevent sarcopenia and other age-related conditions. With ongoing research, we can build a healthier future for the world's aging population.

—*Chad Cox*

PART I

HOW NUTRITION AND OTHER FACTORS RELATE TO SARCOPENIA

Nutrition and Sarcopenia: A Review of the Evidence and Implications for Preventive Strategies

SIBN ROBINSON, CYRUS COOPER, AND AVAN AIHIE SAYER

1.1 INTRODUCTION

Sarcopenia is the loss of muscle mass and strength that occurs with advancing age [1]. Although definitions (and therefore estimates of prevalence) vary, it is widely recognised as a common condition among older adults, and one that is associated with huge personal and financial costs [1, 2]. Declining muscle mass and strength are expected components of ageing. However, the rate of decline differs across the population [1, 3], suggesting that modifiable behavioural factors such as diet and lifestyle may be important influences on muscle function in older age. This paper considers the evidence that links diet to muscle mass and strength, and implications for strategies to prevent or delay sarcopenia in older age.

Nutrition and Sarcopenia: A Review of the Evidence and Implications for Preventive Strategies. © Robinson S, Cooper C, and Sayer AA. Journal of Aging Research **2012** (2012). http://dx.doi. org/10.1155/2012/510801. Licensed under Creative Commons 3.0 Unported License, http://creative-commons.org/licenses/by/3.0/.

1.2 NUTRITION AND AGEING

Food intake falls by around 25% between 40 and 70 years of age [4]. In comparison with younger ages, older adults eat more slowly, they are less hungry and thirsty, consume smaller meals, and they snack less [4]. The mechanisms for the "anorexia of ageing" are not fully understood but there may be a range of physiological, psychological, and social factors that influence appetite and food consumption, including loss of taste and olfaction, increased sensitivity to the satiating effects of meals, chewing difficulties, and impaired gut function [4, 5]. The negative consequences of these changes are compounded by the effects of functional impairments that impact on the ability to access and prepare food, psychological problems such as depression and dementia, as well as the social effects of living and eating alone. Low food intakes and monotonous diets put older people at risk of having inadequate nutrient intakes [6]. Thus in a vicious cycle, declining muscle strength and physical capability in older age may increase the risk of poor nutrition, whilst poor nutrition may contribute to further declines in physical capability.

The exact estimates of the prevalence of poor nutrition may differ according to the definitions used, but studies of community-dwelling adults consistently suggest that it is common in older age. For example in the National Diet and Nutrition Survey in the UK, 14% of older men and women living in the community, and 21% of those living in institutions, were at medium or high risk of undernutrition [7]. Estimates of the prevalence of undernutrition in older patients admitted to hospital are even greater, ranging up to 72% [8, 9]. These figures are clearly substantial and indicate that there are significant numbers of older adults living in developed settings who currently have less than optimal nutrition.

1.3 IS DIET A MODIFIABLE INFLUENCE ON SARCOPENIA?

There are two consequences of declining food intakes in older age that could be important for muscle mass and strength. Firstly, lower energy intakes, if not matched by lower levels of energy expenditure, lead to weight loss, including a loss of muscle mass [4]. Secondly, as older people consume smaller amounts of food, it may become more challenging for them to meet their nutrient needs—particularly for micronutrients. For older people with low food intakes, this highlights the importance of having diets of adequate quality. Although the importance of adequate nutrition has been recognised for a long time, its contribution to muscle mass and strength has not been studied extensively and much of the research in this area is relatively new [10]. A number of interventions have been studied, ranging from provision of nutritional support [11], to supplementation with specific nutrients [12, 13]. The nutrients that have been most consistently linked to sarcopenia and frailty in older adults are vitamin D, protein, and a number of antioxidant nutrients, that include carotenoids, selenium, and vitamins E and C [10]. However, there is also some evidence that variations in long-chain polyunsaturated fatty acid status may have important effects on muscle strength in older people [13].

1.3.1 PROTEIN

Protein is considered a key nutrient in older age [14]. Dietary protein provides amino acids that are needed for the synthesis of muscle protein, and importantly, absorbed amino acids have a stimulatory effect on muscle protein synthesis after feeding [15]. There is some evidence that the synthetic response to amino acid intake may be blunted in older people, particularly at low intakes [14], and when protein is consumed together with carbohydrate [16]. Recommended protein intakes may, therefore, need to be raised in older people in order to maintain nitrogen balance and to protect them from sarcopenic muscle loss [14].

Whilst there is currently no consensus on the degree to which dietary protein requirements change in older age, there is important observational evidence that an insufficient protein intake may be an important contributor to impaired physical function. For example, in the US Health, Aging and Body Composition Study, a greater loss of lean mass over 3 years, assessed using dual-energy X-ray absorptiometry, was found among older community-dwelling men and women who had low energy-adjusted protein intakes at baseline [17]. The differences were substantial, such that the participants with protein intakes in the top fifth of the distribution lost 40% less lean mass over the follow-up period when compared with those in bottom fifth. Protein and/or amino acid supplementation should, therefore, have the potential to slow sarcopenic muscle loss. However, whilst amino acid supplementation has been shown to increase lean mass and improve physical function [18], other trials have not been successful [16, 19]. Further work, including longer-term trials, is needed to define optimal protein intakes in older age [16].

1.3.2 VITAMIN D

An association between vitamin-D-deficient osteomalacia and myopathy has been recognised for many years [20], but the role of vitamin D, and the extent to which it has direct effects on normal muscle strength and physical function remains controversial [21]. The potential mechanisms that link vitamin D status to muscle function are complex and include both genomic and nongenomic roles [20, 22]. The vitamin D receptor (VDR) has been isolated from skeletal muscle, indicating that it is a target organ [20], and polymorphisms of the VDR have been shown to be related to differences in muscle strength [23]. At the genomic level, binding of the biologically active form of the vitamin (1,25-dihydroxyvitamin D) results in enhanced transcription of a range of proteins, including those involved in calcium metabolism [20]. The nongenomic actions of vitamin D are currently less well understood [22].

Much of the epidemiological literature is consistent with the possibility that there are direct effects of vitamin D on muscle strength. For example, among men and women aged 60 years and older in NHANES III, low vitamin D status (serum 25-hydroxyvitamin D < 15 ng mL^{-1}) was associated with a fourfold increase in risk of frailty [24], and in a meta-analysis of supplementation studies of older adults, Bischoff-Ferrari et al. [12] showed that supplemental vitamin D (700–1000 IU per day) reduced the risk of falling by 19%. However, the evidence is not always consistent as some observational studies find no association between vitamin D status and physical function, and supplementation studies have not always resulted in measurable improvements in function [21]. In a review of published studies, Annweiler and colleagues [21] discuss the reasons for the divergence in study findings, some of which may be due to methodological differences, including a lack of consideration of confounding influences in some studies. Further evidence is needed, particularly as vitamin D insufficiency is common among older adults [24].

1.3.3 ANTIOXIDANT NUTRIENTS

There is increasing interest in the role of oxidative stress in aetiology of sarcopenia, and markers of oxidative damage have been shown to predict impairments in physical function in older adults [25]. Damage to biomolecules such as DNA, lipid, and proteins may occur when reactive oxygen species (ROS) are present in cells in excess. The actions of ROS are normally counterbalanced by antioxidant defence mechanisms that include the enzymes superoxide dismutase and glutathione peroxidase, as well exogenous antioxidants derived from the diet, such as selenium, carotenoids, tocopherols, flavonoids, and other plant polyphenols [15, 25]. In older age, an accumulation of ROS may lead to oxidative damage and contribute to losses of muscle mass and strength [15].

A number of observational studies have shown positive associations between higher antioxidant status and measures of physical function [10]. Importantly these associations are seen both in cross-sectional analyses and in longitudinal studies, such that poor status is predictive of decline in function. The observed effects are striking. For example, among older men and women in the InCHIANTI study, higher plasma carotenoid concentrations were associated with a lower risk of developing a severe walking disability over a follow-up period of 6 years; after taking account of confounders that included level of physical activity and other morbidity, the odds ratio was 0.44 (95% CI 0.27–0.74) [26]. Inverse associations have also been described for vitamin E and selenium status and risk of impaired physical function [10]. There have been few studies of older adults to determine how antioxidant supplementation affects muscle strength, and the benefits of supplementation remain uncertain [27]. Since ROS have both physiological and pathological roles, interventions based on simple suppression of their activities may be unlikely to improve age-related declines in muscle mass and function [28]. However, low antioxidant intakes and status are common [6, 29], and this remains an important question to be addressed.

1.3.4 LONG-CHAIN POLYUNSATURATED FATTY ACIDS (LCPUFAS)

Sarcopenia is increasingly recognised as an inflammatory state driven by cytokines and oxidative stress [30]. Since eicosanoids derived from 20-carbon polyunsaturated fatty acids are among the mediators and regulators of inflammation [13], this raises the possibility that variations in intake of n-3 and n-6 LCPUFAs, and their balance in the diet, could be of importance. In particular, n-3 LPUFAs have the potential to be potent anti-inflammatory agents [13]. There is some observational evidence to support an effect of n-3 LCPUFA status on muscle function, as higher grip strength was found in older men and women who had greater consumption of oily fish [31]—one of the richest sources of n-3 LCPUFAs in the UK diet. Consistent with this finding, a number of studies of patients with rheumatoid arthritis have shown that supplementation with fish oil resulted in improved grip strength [13]. In a recent randomised controlled trial, supplementation of older adults with n-3 LCPUFA (eicosapentaenoic and docosahexaenoic acids) resulted in an enhanced anabolic response to amino acid and insulin infusion. Whilst these novel data suggest that the stimulation of muscle protein synthesis by n-3 LCPUFA supplementation could be useful for the prevention and treatment of sarcopenia [32], further evidence is needed to establish the therapeutic potential of n-3 LCPUFAs in inflammatory conditions [13].

1.3.5 FOODS AND DIETARY PATTERNS

One problem with the existing evidence base is that dietary components are often highly correlated with each other. This may help to explain why the effects of supplementation with single nutrients may be less than that predicted by the observational evidence. It also means that from observational data it may be difficult to understand the relative importance of the influences of different nutrients on sarcopenia. For example, whilst an antioxidant nutrient such as β-carotene may be causally related to variations in physical function, it may also be acting as a marker of other components of fruit and vegetables. In turn, since diets are patterned, high fruit and vegetable consumption may be indicators of other dietary differences which could be important for muscle function, such as greater consumption of oily fish and higher intakes of vitamin D and n-3 LCPUFAs [33]. The cumulative effects of nutrient deficiencies have been described by Semba et al. [34], in which he estimated that each additional nutrient deficiency raised the risk of frailty in older women by almost 10%. This emphasises the importance of the quality of diets of older adults, as well as the quantity of food consumed, to ensure that intakes of a range of nutrients are sufficient.

Compared with the evidence that links variations in nutrient intake and status to physical function, much less is known about the influence of dietary patterns and dietary quality in older age. "Healthy" diets, characterised by greater fruit and vegetable consumption, wholemeal cereals, and oily fish, have been shown to be associated with greater muscle strength in older adults [31]. Data from studies of younger adults appear to be consistent with this finding. For example, among women aged 42–52 years, "unhealthy" diets, characterised by higher saturated fat intakes and lower fruit and vegetable consumption, were associated with greater functional limitations over a 4-year follow-up period [35]. Benefits of healthier diets and greater fruit and vegetable consumption on physical function in midlife have also been described in women in the Whitehall study [36], and in men and women in the Atherosclerosis Risk in Communities Study [37]. Intervention studies that take a food-based or "whole diet" approach are likely to change intakes of a range of nutrients and, therefore, have the potential to be more effective than single nutrient supplementation studies in preventing age-related losses in muscle mass and strength.

1.3.6 DIET AND EXERCISE

Resistance exercise training interventions have been shown to be effective in increasing muscle strength and improving physical function in older adults [38]. A further issue in understanding a possible protective role for diet in sarcopenia is, therefore, the potential for interactions between diet and exercise, and the extent to which interventions that combine supplementation and exercise training may be more effective than changing nutrient intake alone. The interactive effects of diet and exercise on physical function have been studied most extensively in relation to protein/amino acid supplementation. For example, whilst consumption of a high protein meal has been shown to increase muscle protein synthesis in older adults by ~50%, combining a high protein meal with resistance exercise increases synthesis more than 100% [39]. However, a number of studies of older adults have failed to show additional benefits of protein/amino acid supplementation on the skeletal muscle response to prolonged resistance exercise training [15, 40], and the implications for long-term effects of combined exercise training and high protein intakes are, therefore, not clear [16]. Current findings point to the need for further research—particularly to address the effects of differing quantity and timing of supplementation [39, 40]. At present we have limited insights into the combined effects of vitamin D supplementation and resistance exercise on muscle strength and function [41].

1.4 LIFELONG NUTRITION AND SARCOPENIA

An important limitation to the current evidence base that links nutrition to sarcopenia is that much of the observational data are from cross-sectional studies. Aside from methodological considerations of studying older adults who may have a number of comorbidities, this raises particular issues that may limit our understanding of the potential importance of the role of nutrition in the loss of muscle mass and strength with age.

Firstly, the health of older people is influenced by events throughout their lives [10], and achievement of optimal function may, therefore, depend on lifelong exposure to a healthy diet and lifestyle. Although there is evidence that healthier eating behaviours are reasonably stable in adult life [42], little is known of changes in dietary habits in older people, at a time when morbidity-related dietary advice is available, and lifestyle may be changing rapidly. The influence of lifelong nutrition on age-related changes in muscle mass and strength has been little studied, but in terms of interventions to delay or prevent sarcopenia in older age, there may be key opportunities earlier in the lifecourse that need to be recognised.

A second consideration is that muscle mass and strength achieved in later life are not only determined by the rate of muscle loss, but also reflect the peak attained in early life (Figure 1, [43]). Thus, factors that influence growth, such as variations in early nutrition, may contribute to muscle mass and strength in older age.

A key finding, that highlights the importance of lifecourse influences, is that low weight at birth predicts lower muscle mass and strength in adult life. This is a consistent finding across a number of studies [44]. Although little is currently known about the influence of diet in early life on sarcopenia, recent studies of adolescents have provided evidence of nutrient effects on muscle mass and function earlier in the lifecourse. Consistent with studies of older adults, low vitamin D status has been shown to be associated both with lower grip strength and with poorer muscle power and velocity [45, 46]. However, randomized controlled trials of vitamin D supplementation of adolescents have had mixed results. Among premenarcheal girls who were supplemented with vitamin D over 1 year, there were graded increases in lean mass, although supplementation did not result in measurable differences in grip strength [47]. In contrast, vitamin D supplementation of adolescent boys and postmenarcheal girls has not been shown to be effective in increasing lean mass or muscle strength or power [47, 48]. Ward et al. [48] conclude that earlier interventions, before the period of peak muscle mass accretion, may be needed to improve muscle function and physical performance.

To date, few studies have examined the role of diet in early childhood in the acquisition of muscle mass and effects on later function, although there is some evidence that it could be important. For example, the risk of frailty has been shown to be greater in older adults who grew up in impoverished circumstances, and who experienced hunger in childhood [49]. However, animal models suggest that nutrition even earlier in life may be key, as muscle growth in the neonatal period is highly sensitive to variations in nutrient intake [50]. In two recent studies, the role of variations in infant diet has been addressed, but with differing results. Among children in the ALSPAC study, duration of breastfeeding was not associated with physical work capacity assessed at the age 9 years [51], whilst in adolescents studied in the HELENA cohort, longer duration of breastfeeding was associated with measurable differences in physical performance–particularly in lower body explosive strength [52]. Consistent with these latter findings, longer duration of breastfeeding and greater compliance with infant feeding guidance has been shown to be associated with greater lean mass in later childhood [53]. Dietary patterns "track" across childhood [54], and this may simply reflect continuing benefits of healthier diets. However, it does suggest that variations in early postnatal diet could have implications for muscle function in later life.

We currently know little about the contribution of nutrition across the lifecourse to muscle mass and strength in adult life, and further work is needed to understand how early nutrition influences the acquisition of peak muscle mass, and the role played by nutrition in the trajectory of age-related losses in muscle function. Taking a lifecourse approach to understanding the links between nutrition and muscle mass and function in older age could change dietary strategies to prevent sarcopenia in the future.

1.5 CONCLUSION

To develop strategies to prevent or delay sarcopenia, a better understanding is needed of the lifestyle factors that influence the rate of decline of muscle mass and strength in older age, and the mechanisms involved. Existing evidence indicates the potential importance of diets of adequate quantity and quality, to ensure sufficient intakes of protein, vitamin D, and antioxidant nutrients. Although much of this evidence is observational and the mechanisms are not fully understood, the high prevalence of low nutrient intakes and poor status among older adults make this a current concern. However, muscle mass and strength achieved in later life are not only determined by the rate of muscle loss, but also reflect the peak attained earlier in life, and efforts to prevent sarcopenia also need to recognise the potential effectiveness of interventions earlier in the lifecourse. Optimising diet and nutrition throughout life may be key to preventing sarcopenia and promoting physical capability in older age.

FIGURE 1: A lifecourse model of sarcopenia (from [43]).

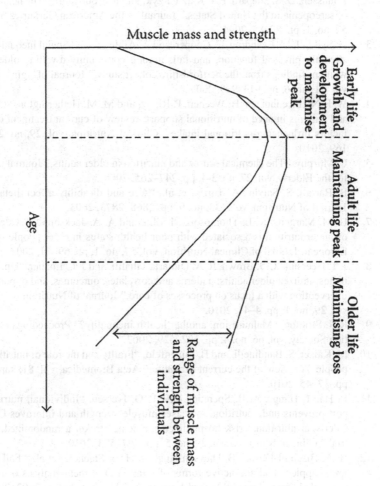

REFERENCES

1.   A. J. Cruz-Jentoft, J. P. Baeyens, J. M. Bauer et al., "Sarcopenia: European consensus on definition and diagnosis," Age and Ageing, vol. 39, no. 4, Article ID afq034, pp. 412–423, 2010.

2.   I. Janssen, D. S. Shepard, P. T. Katzmarzyk, and R. Roubenoff, "The healthcare costs of sarcopenia in the United States," Journal of the American Geriatrics Society, vol. 52, no. 1, pp. 80–85, 2004.

3.   H. Syddall, M. Evandrou, C. Cooper, and A. Aihie Sayer, "Social inequalities in grip strength, physical function, and falls among community dwelling older men and women: findings from the hertfordshire cohort study," Journal of Aging and Health, vol. 21, no. 6, pp. 913–939, 2009.

4.   W. F. Nieuwenhuizen, H. Weenen, P. Rigby, and M. M. Hetherington, "Older adults and patients in need of nutritional support: review of current treatment options and factors influencing nutritional intake," Clinical Nutrition, vol. 29, no. 2, pp. 160–169, 2010.

5.   C. Murphy, "The chemical senses and nutrition in older adults," Journal of Nutrition for the Elderly, vol. 27, no. 3-4, pp. 247–265, 2008.

6.   B. Bartali, S. Salvini, A. Turrini et al., "Age and disability affect dietary intake," Journal of Nutrition, vol. 133, no. 9, pp. 2868–2873, 2003.

7.   B. M. Margetts, R. L. Thompson, M. Elia, and A. A. Jackson, "Prevalence of risk of undernutrition is associated with poor health status in older people in the UK," European Journal of Clinical Nutrition, vol. 57, no. 1, pp. 69–74, 2003.

8.   J. T. Heersink, C. J. Brown, R. A. Dimaria-Ghalili, and J. L. Locher, "Undernutrition in hospitalized older adults: patterns and correlates, outcomes, and opportunities for intervention with a focus on processes of care," Journal of Nutrition for the Elderly, vol. 29, no. 1, pp. 4–41, 2010.

9.   R. J. Stratton, "Malnutrition: another health inequality?" Proceedings of the Nutrition Society, vol. 66, no. 4, pp. 522–529, 2007.

10.   M. Kaiser, S. Bandinelli, and B. Lunenfeld, "Frailty and the role of nutrition in older people. A review of the current literature," Acta Biomedica, vol. 81, supplement 1, pp. 37–45, 2010.

11.   L. Ha, T. Hauge, A. B. Spenning, and P. O. Iversen, "Individual, nutritional support prevents undernutrition, increases muscle strength and improves QoL among elderly at nutritional risk hospitalized for acute stroke: a randomized, controlled trial," Clinical Nutrition, vol. 29, no. 5, pp. 567–573, 2010.

12.   H. A. Bischoff-Ferrari, B. Dawson-Hughes, H. B. Staehelin et al., "Fall prevention with supplemental and active forms of vitamin D: a meta-analysis of randomised controlled trials," British Medical Journal, vol. 339, Article ID b3692, 2009.

13.   P. C. Calder, "n-3 Polyunsaturated fatty acids, inflammation, and inflammatory diseases," American Journal of Clinical Nutrition, vol. 83, supplement 6, pp. 1505S–1519S, 2006.

14.   R. R. Wolfe, S. L. Miller, and K. B. Miller, "Optimal protein intake in the elderly," Clinical Nutrition, vol. 27, no. 5, pp. 675–684, 2008.

15. J. S. Kim, J. M. Wilson, and S. R. Lee, "Dietary implications on mechanisms of sarcopenia: roles of protein, amino acids and antioxidants," Journal of Nutritional Biochemistry, vol. 21, no. 1, pp. 1–13, 2010.

16. D. Paddon-Jones and B. B. Rasmussen, "Dietary protein recommendations and the prevention of sarcopenia," Current Opinion in Clinical Nutrition and Metabolic Care, vol. 12, no. 1, pp. 86–90, 2009.

17. D. K. Houston, B. J. Nicklas, J. Ding et al., "Dietary protein intake is associated with lean mass change in older, community-dwelling adults: the Health, Aging, and Body Composition (Health ABC) study," American Journal of Clinical Nutrition, vol. 87, no. 1, pp. 150–155, 2008.

18. E. Børsheim, Q. T. Bui, S. Tissier, H. Kobayashi, A. A. Ferrando, and R. R. Wolfe, "Effect of amino acid supplementation on muscle mass, strength and physical function in elderly," Clinical Nutrition, vol. 27, no. 2, pp. 189–195, 2008.

19. A. C. Milne, J. Potter, A. Vivanti, and A. Avenell, "Protein and energy supplementation in elderly people at risk from malnutrition," Cochrane Database of Systematic Reviews, no. 2, Article ID CD003288, 2009.

20. B. Hamilton, "Vitamin D and human skeletal muscle," Scandinavian Journal of Medicine and Science in Sports, vol. 20, no. 2, pp. 182–190, 2010.

21. C. Annweiler, A. M. Schott, G. Berrut, B. Fantino, and O. Beauchet, "Vitamin D-related changes in physical performance: a systematic review," Journal of Nutrition, Health and Aging, vol. 13, no. 10, pp. 893–898, 2009.

22. L. Ceglia, "Vitamin D and its role in skeletal muscle," Current Opinion in Clinical Nutrition and Metabolic Care, vol. 12, no. 6, pp. 628–633, 2009.

23. P. Geusens, C. Vandevyver, J. Vanhoof, J. J. Cassiman, S. Boonen, and J. Raus, "Quadriceps and grip strength are related to vitamin D receptor genotype in elderly nonobese women," Journal of Bone and Mineral Research, vol. 12, no. 12, pp. 2082–2088, 1997.

24. E. R. Wilhelm-Leen, Y. N. Hall, I. H. de boer, and G. M. Chertow, "Vitamin D deficiency and frailty in older Americans," Journal of Internal Medicine, vol. 268, no. 2, pp. 171–180, 2010.

25. R. D. Semba, L. Ferrucci, K. Sun et al., "Oxidative stress and severe walking disability among older women," The American Journal of Medicine, vol. 120, no. 12, pp. 1084–1089, 2007.

26. F. Lauretani, R. D. Semba, S. Bandinelli et al., "Carotenoids as protection against disability in older persons," Rejuvenation Research, vol. 11, no. 3, pp. 557–563, 2008.

27. D. Fusco, G. Colloca, M. R. Lo Monaco, and M. Cesari, "Effects of antioxidant supplementation on the aging process," Clinical Interventions in Aging, vol. 2, no. 3, pp. 377–387, 2007.

28. M. J. Jackson, "Strategies for reducing oxidative damage in ageing skeletal muscle," Advanced Drug Delivery Reviews, vol. 61, no. 14, pp. 1363–1368, 2009.

29. J. P. Stimpson, A. C. Nash, H. Ju, and K. Eschbach, "Neighborhood deprivation is associated with lower levels of serum carotenoids among adults participating in the third national health and nutrition examination survey," Journal of the American Dietetic Association, vol. 107, no. 11, pp. 1895–1902, 2007.

30. G. L. Jensen, "Inflammation: roles in aging and sarcopenia," Journal of Parenteral and Enteral Nutrition, vol. 32, no. 6, pp. 656–659, 2008.

31. S. M. Robinson, K. A. Jameson, S. F. Batelaan et al., "Diet and its relationship with grip strength in community-dwelling older men and women: the Hertfordshire cohort study," Journal of the American Geriatrics Society, vol. 56, no. 1, pp. 84–90, 2008.

32. G. I. Smith, P. Atherton, D. N. Reeds et al., "Dietary omega-3 fatty acid supplementation increases the rate of muscle protein synthesis in older adults: a randomized controlled trial," American Journal of Clinical Nutrition, vol. 93, no. 2, pp. 402–412, 2011. View at Publisher · View at Google Scholar

33. S. Robinson, H. Syddall, K. Jameson et al., "Current patterns of diet in community-dwelling older men and women: results from the Hertfordshire cohort study," Age and Ageing, vol. 38, no. 5, pp. 594–599, 2009.

34. R. D. Semba, B. Bartali, J. Zhou, C. Blaum, C. W. Ko, and L. P. Fried, "Low serum micronutrient concentrations predict frailty among older women living in the community," The Journals of Gerontology Series A, vol. 61, no. 6, pp. 594–599, 2006.

35. K. M. Tomey, M. R. Sowers, C. Crandall, J. Johnston, M. Jannausch, and M. Yosef, "Dietary intake related to prevalent functional limitations in midlife women," American Journal of Epidemiology, vol. 167, no. 8, pp. 935–943, 2008.

36. M. Stafford, H. Hemingway, S. A. Stansfeld, E. Brunner, and M. Marmot, "Behavioural and biological correlates of physical functioning in middle aged office workers: the UK whitehall II study," Journal of Epidemiology and Community Health, vol. 52, no. 6, pp. 353–358, 1998.

37. D. K. Houston, J. Stevens, J. Cai, and P. S. Haines, "Dairy, fruit, and vegetable intakes and functional limitations and disability in a biracial cohort: the atherosclerosis risk in communities study," American Journal of Clinical Nutrition, vol. 81, no. 2, pp. 515–522, 2005.

38. C. J. Liu and N. K. Latham, "Progressive resistance strength training for improving physical function in older adults," Cochrane Database of Systematic Reviews, no. 3, Article ID CD002759, 2009.

39. T. B. Symons, M. Sheffield-Moore, M. M. Mamerow, R. R. Wolfe, and D. Paddon-Jones, "The anabolic response to resistance exercise and a protein-rich meal is not diminished by age," Journal of Nutrition, Health and Aging, vol. 15, no. 5, pp. 376–381, 2010. View at Publisher · View at Google Scholar

40. R. Koopman, "Exercise and protein nutrition: dietary protein and exercise training in ageing," Proceedings of the Nutrition Society, vol. 70, no. 01, pp. 104–113, 2010.

41. D. Bunout, G. Barrera, L. Leiva et al., "Effects of vitamin D supplementation and exercise training on physical performance in Chilean vitamin D deficient elderly subjects," Experimental Gerontology, vol. 41, no. 8, pp. 746–752, 2006.

42. G. D. Mishra, S. A. McNaughton, G. D. Bramwell, and M. E. J. Wadsworth, "Longitudinal changes in dietary patterns during adult life," British Journal of Nutrition, vol. 96, no. 4, pp. 735–744, 2006.

43. A. A. Sayer, H. Syddall, H. Martin, H. Patel, D. Baylis, and C. Cooper, "The developmental origins of sarcopenia," Journal of Nutrition, Health and Aging, vol. 12, no. 7, pp. 427–431, 2008.

44. A. A. Sayer, "Sarcopenia," British Medical Journal, vol. 341, Article ID c4097, 2010. View at Google Scholar

45. L. H. Foo, Q. Zhang, K. Zhu et al., "Low vitamin D status has an adverse influence on bone mass, bone turnover, and muscle strength in Chinese adolescent girls," Journal of Nutrition, vol. 139, no. 5, pp. 1002–1007, 2009.

46. K. A. Ward, G. Das, J. L. Berry et al., "Vitamin D status and muscle function in postmenarchal adolescent girls," Journal of Clinical Endocrinology and Metabolism, vol. 94, no. 2, pp. 559–563, 2009.

47. G. E. H. Fuleihan, M. Nabulsi, H. Tamim et al., "Effect of vitamin D replacement on musculoskeletal parameters in school children: a randomized controlled trial," Journal of Clinical Endocrinology and Metabolism, vol. 91, no. 2, pp. 405–412, 2006.

48. K. A. Ward, G. Das, S. A. Roberts et al., "A randomized, controlled trial of vitamin D supplementation upon musculoskeletal health in postmenarchal females," Journal of Clinical Endocrinology and Metabolism, vol. 95, no. 10, pp. 4643–4651, 2010.

49. B. E. Alvarado, M. V. Zunzunegui, F. Béland, and J. M. Bamvita, "Life course social and health conditions linked to frailty in latin american older men and women," The Journals of Gerontology Series A, vol. 63, no. 12, pp. 1399–1406, 2008.

50. T. A. Davis and M. L. Fiorotto, "Regulation of muscle growth in neonates," Current Opinion in Clinical Nutrition and Metabolic Care, vol. 12, no. 1, pp. 78–85, 2009.

51. D. A. Lawlor, A. R. Cooper, C. Bain et al., "Associations of birth size and duration of breast feeding with cardiorespiratory fitness in childhood: findings from the Avon Longitudinal Study of Parents and Children (ALSPAC)," European Journal of Epidemiology, vol. 23, no. 6, pp. 411–422, 2008.

52. E. G. Artero, F. B. Ortega, V. Espana-Romero et al., "Longer breastfeeding is associated with increased lower body explosive strength during adolescence," Journal of Nutrition, vol. 140, no. 11, pp. 1989–1995, 2010.

53. S. M. Robinson, L. D. Marriott, S. R. Crozier et al., "Variations in infant feeding practice are associated with body composition in childhood: a prospective cohort study," Journal of Clinical Endocrinology and Metabolism, vol. 94, no. 8, pp. 2799–2805, 2009.

54. K. Northstone and P. M. Emmett, "Are dietary patterns stable throughout early and mid-childhood? A birth cohort study," British Journal of Nutrition, vol. 100, no. 5, pp. 1069–1076, 2008.

CHAPTER 2

Malnutrition-Sarcopenia Syndrome: Is This the Future of Nutrition Screening and Assessment for Older Adults?

MAURITS F. J. VANDEWOUDE, CAROLYN J. ALISH, ABBY C. SAUER, AND REFAAT A. HEGAZI

2.1 INTRODUCTION

Historically, malnutrition has been defined as a condition of an imbalance of energy, protein, and other nutrients that cause measurable negative effects on body composition, physical function, and clinical outcomes [1]. Typical measures that clinicians use to screen and assess for malnutrition or the risk for malnutrition include dietary or nutrient intake, changes in body weight, and laboratory values [2]. A new definition of malnutrition has recently been proposed by an International Guideline Consensus Committee, integrating the acuity of the associated disease and inflammation [3]. The committee specified three subtypes of malnutrition using an etiology-based terminology to assist clinicians to make a nutrition diagnosis in clinical practice set-

Malnutrition-Sarcopenia Syndrome: Is This the Future of Nutrition Screening and Assessment for Older Adults? © *Vandewoude MFJ, Alish CJ, Sauer AC, and Hegazi RA.* Journal of Aging Research *2012 (2012). http://dx.doi.org/10.1155/2012/651570. Licensed under Creative Commons 3.0 Unported License, http://creativecommons.org/licenses/by/3.0/.*

tings: (1) starvation-related without inflammation, (2) chronic disease or conditions that impose sustained mild-to-moderate inflammation (e.g., sarcopenic obesity, organ failure, and pancreatic cancer), and (3) acute disease or injury states, when inflammatory response is marked [3].

One critical clinical aspect often not assessed in nutrition screening or assessment is lean body mass or muscle mass loss. Lean body mass (LBM) is defined as that portion of the body mass that is everything but fat and includes water, mineral, muscle, and other protein-rich structures (e.g., enzymes, viscera, red cells, and connective tissues) [4]. Skeletal muscle mass constitutes the majority of LBM and provides strength, mobility, and balance [5]. Muscle mass also plays a critical role in whole-body protein metabolism and impacts quality of life in patients with chronic diseases [6]. The balance between muscle protein anabolism and catabolism is vitally important to maintaining skeletal muscle mass, particularly in older adults who lose muscle mass as a consequence of aging and/or illness [6, 7]. It was not until 1989, when Irwin Rosenberg introduced the term sarcopenia [8]. The European Working Group on Sarcopenia in Older People (EWGSOP) defines sarcopenia as an age-related loss of muscle mass, combined with loss of strength, functionality, or both (Figure 1) [9]. The working group also proposed a diagnostic algorithm for sarcopenia that is based on the presence of low muscle mass plus either low muscle strength (e.g., low handgrip strength) or low physical performance (e.g., 4 meter walking speed).

Sarcopenia is further classified into either primary or secondary categories. Primary sarcopenia, when no specific etiologic cause can be identified, is progressive and associated with the impact of aging: a reduction in motor neurons, alterations in skeletal muscle tissue including mitochondrial dysfunction, changes in the hormonal milieu (e.g., insulin resistance and a reduction in insulin-like growth factor-1 and an increase in proinflammatory cytokines, such as tumor necrosis factor α and interleukin 6. Next to the intrinsic, age-related processes, a multitude of extrinsic and behavioral factors can aggravate the development and/or progression of sarcopenia, leading to secondary sarcopenia, such as disuse and lack of physical activity, malnutrition, chronic inflammation, and comorbidity. As such, sarcopenia can be thought of as both a process and an outcome. Sarcopenia as a condition is a major cause of frailty and disability in older adults; as an active process, it is present in every person reaching adult life [9].

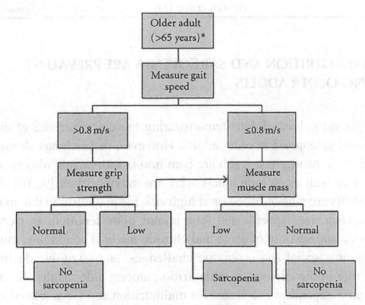

*Comorbidity and individual circumstances that may explain each finding must be considered
*This algorithm can also be applied to younger individuals at risk

FIGURE 1: EUGMS working group-suggested algorithm for sarcopenia case finding in older individuals (used with permission) [9].

TABLE 1: Prevalence of malnutrition and sarcopenia in older adults across clinical settings.

Patient Population/Setting	Malnutrition	Sarcopenia
Hospital/Acute Care	56% [11]	
	23% [15]	
Long Term Care	21% [15]	32.8% [21]
	2–9% [15]	25% [20]
Community	1–10% [15]	37–61% (class I sarcopenia) [22]
	(41–48% at risk) [14]	5–11% (class II sarcopenia) [22]

2.2 MALNUTRITION AND SARCOPENIA ARE PREVALENT AMONG OLDER ADULTS

There are no published data demonstrating the co-occurrence of malnutrition and sarcopenia in older adults. However, research has shown that reductions in handgrip strength are common in individuals who have sarcopenia as well as in individuals who are malnourished [9, 10]. Many older adults are malnourished or at high risk for malnutrition due to many factors. Decreased appetite and food intake, poor dentition, an increased frequency and severity of acute and chronic medical conditions, multiple medications, social and economic challenges, and cognitive decline all play a role in the etiology of malnutrition among older adults. Advanced age is an independent risk factor for malnutrition and is associated with a lower body weight, body mass index (BMI), and serum albumin [11–13].

The prevalence of malnutrition is greater among older adults in health care settings than in the community (Table 1). In hospital settings, malnutrition among older patients is approximately 56% [11]. For older adults living in the community, the prevalence of malnutrition ranges from 1 to 10%, while 41–48% are at risk for malnutrition [14]. Consistently, in a large international study, malnutrition was present in 2% of those living in the community, 9% of outpatients and home care combined, 23% in the hospital, 21% in institutions, and 15% in those with cognitive impairment [15].

Age-associated loss of muscle mass is characterized by a 3–8% decline per decade after the age of 30 years with further decline in adults 60 years of age and older [7, 16]. With aging, the loss of muscle mass is accompanied by an increase in body fat [17]. On average, adults can experience annual losses of 0.23 kg of muscle and gains of 0.45 kg of fat between 30 and 60 years of age [18]. Acute illness and injury can accelerate age-related changes in body composition. Studies demonstrate that, following injury, patients can lose 5-6% of their total body weight—most of which is muscle mass —and gain between 4 and 11% in fat mass within 12 months, with the majority of LBM loss occurring within the first two to four months after the injury [19].

Using the EWGSOP diagnostic criteria, Landi et al. evaluated the baseline data of adults who were 80 years of age or older (n = 260) from the il-SIRENTE study. The results of the study indicated that sarcopenia is prevalent among community-dwelling older adults with no differences based on gender (25%) [20]. Using the same diagnostic algorithm, Landi et al. demonstrated that the prevalence of sarcopenia is slightly higher (32.8%) among older adults in long-term care settings and was higher among male residents (68%) than among female residents (21%) [21].

Reductions in muscle mass among older adults are common and can contribute to functional impairment, disability, increased risk for falls [20], lowered quality of life, and increased risk for mortality. For instance, using cross-sectional body composition data from the Third National Health and Nutrition Examination Survey, Janssen and colleagues determined the prevalence of class I sarcopenia (skeletal muscle mass index within one to two standard deviations below sex-specific values for young adults) and class II sarcopenia (skeletal muscle mass index two standard deviations of young adult values) in adults age 50 years of older. Among adults 50 years of age and older, the prevalence of class I sarcopenia was estimated between 37 and 47% in men and 50–61% in women, and a prevalence of class II sarcopenia between 5–7% in men and 7–11% in women. The prevalence of sarcopenia increased from the third to sixth decades and remained constant among subjects 70 years of age and older, while the prevalence of class I (59% versus 45%) and class II (10% versus 7%) was greater in women ≥50 years of age than men (P < 0.001). Moreover, functional impairment was 2-3-fold higher in older subjects with class II sarcopenia than those with normal muscle mass, even after adjusting for age, race, BMI, health behaviors, and co-morbidities [22].

2.3 MALNUTRITION SARCOPENIA SYNDROME
AND CLINICAL OUTCOMES

In many patient populations, malnutrition and sarcopenia are present in parallel and manifest clinically through a combination of decreased nutrient intake, decreased body weight, along with a decrease in muscle mass, strength, and/or physical function. This leads us to coin the proposed clinical syndrome of Malnutrition-Sarcopenia Syndrome (MSS). MSS is the clinical presentation of both malnutrition and accelerated age-associated loss of lean body mass, strength, and/or physical performance. Malnutrition and sarcopenia are each independently associated with negative health consequences that impact older adults across health care settings. Patients with malnutrition and/or sarcopenia are at risk of increased morbidity and mortality, decreased quality of life and functioning and increased rehospitalization, length of hospital stay, and healthcare costs. Importantly, malnutrition and sarcopenia are associated with increased mortality [23–27]. Cederholm et al. found significant differences in mortality rates between malnourished patients and well-nourished patients after hospitalization (44% versus 18%, respectively) [23]. In community-dwelling older adults, unintentional weight loss and low BMI are associated with elevated 3-year mortality rates, and older adults reporting unintentional weight loss were 1.67 times more likely to experience mortality than those who reported no weight loss [28]. Newman and colleagues demonstrated that 5% weight loss over a three-year period is a significant and independent predictor of mortality in community-dwelling aging adults [17]. Similarly, Cereda et al. determined that low BMI predicted mortality in older institutionalized adults [29]. Interestingly, a prospective observational cohort of older adults demonstrated that higher lean mass and lean mass index predicted lower mortality with an 85% reduction in risk, suggesting that changes in lean mass and lean mass index, rather than BMI, are better predictors of mortality in older adults and highlighting the role of lean muscle mass loss in defining malnutrition [30]. Sarcopenia has also been linked with increased mortality in various patient populations. Recently, Landi et al. demonstrated that sarcopenia is highly prevalent in older nursing home residents and is associated with a significantly increased risk of all-cause death [31]. Additionally, Bunout et al. evaluated the relationship between the loss of fat-free mass and mortality among aging adults (n = 1413, 74.3 ± 5.6 years of age) and showed that low fat-free mass was significantly associated with mortality among individuals over 74 years of age [32]. Furthermore, results from a recent meta-analysis show that objective measures of physical capability, such as handgrip strength, walking speed, and chair rise, are predictors of all-cause mortality in older community-dwelling populations [33]. The loss of function associated with sarcopenia and malnutrition is a risk factor for negative outcomes.

Malnutrition and sarcopenia are associated with increased morbidity, in particular increased infection and complications rates, including falls [34] and disability [35]. In a study of hospitalized patients, malnutrition was shown to be a significant independent risk factor for nosocomial infections, with infection rates of 4.4% in the well-nourished group, 7.6% in the moderate malnutrition group as compared to 14.6% in the severely malnourished group [36]. In addition, serum albumin levels, age, weight, immunodeficiency, and nutrition risk index score were associated with increased risk of nosocomial infections [36]. Edington et al. determined that malnourished patients experience a two-fold increase in rates of infection as compared to well-nourished patients, indicating that malnutrition is associated with hospital-acquired infections [37]. Similarly, a study of hospitalized older adults demonstrated that patients identified with sarcopenia (detected by dual energy X-ray absorptiometry (DXA)) upon admission were at a greater risk of contracting a nosocomial infection during the first 3 weeks of hospitalization (relative risk of 2.1) [38].

Malnutrition and loss of muscle mass compromise the quality of life and functional capacity of aging adults. Malnutrition is associated with declines in functional capacity in hospitalized patients [39]. Moreover, reduced quality of life has been reported among malnourished patients with a total Mini-Nutritional Assessment (MNA) score <24 [40]. Malnutrition significantly impacts the clinical outcomes of community-dwelling older adults. Specifically, older patients who are malnourished are more likely to be discharged to a residential home and less likely to return home [41, 42] with an increased length of convalescence, greater disability and dependence on walking devices, and loss of muscle strength after hospitalization [43–45]. Other consequences of sarcopenia are persistent sense of fatigue, muscle weakness, increased predisposition to metabolic disorders, and increased risk of falls and fractures [46]. Studies suggest that loss of muscle mass is a predictor of functional decline in independent older adults and those with disability [22, 47] and that age-related loss of muscle mass is directly correlated with loss in strength [48]. Interestingly, Reid et al. showed that lower extremity muscle mass is a strong independent predictor of the level of functional impairment [49]. Loss of strength with aging tends to track with loss of muscle mass in physiological studies although the decline in muscle strength is steeper than the decline in muscle mass [50]. Also, interventions that increase muscle mass do not necessarily increase strength, and changes in strength that occur with resistance training precede measurable changes in muscle mass. Correlations between change in muscle mass and change in strength in older adults are therefore not consistent. Recently, the term "sarcopenia with limited mobility" was proposed as a syndrome that occurs when sarcopenia leads to loss of function and individuals become candidates for therapeutic interventions [51].

In addition to increased risk of infections and functional impairment, malnutrition and loss of muscle mass are also associated with increased hospital length of stay (LOS) [11, 37, 52–55]. Moreover, studies have shown that weight loss and malnutrition are predictors of increased rehospitalization rates in adults [56–59]. In one study, LOS is significantly shorter among well-nourished patients (5.7 days) as compared to malnourished patients (8.9 days) [60]. Leandro-Merhi et al. reported that well-nourished patients are three times more likely to be discharged sooner than patients with varying degrees of malnutrition [55]. Additionally, severely malnourished patients with a BMI < 20 kg/m^2 or weight loss of greater than 10% stayed in the hospital even longer at 18.3 and 17.5 days, respectively [37]. Hospitalization is associated with significant reductions in muscle mass and strength and functional decline in older adults [44, 61–64]. Interestingly, in a large sample of patients, Pichard et al. determined that fat-free mass (FFM) and fat-free mass index are significantly lower among elderly hospitalized patients than their nonhospitalized counterparts. Additionally, 37% of patients hospitalized for just 1-2 days had low FFM, which increased to 55.6% after 12 days of hospitalization [65]. Another study concluded that short-term hospitalization was associated with significant declines in functional capacity and muscle strength, regardless of age or baseline functional status [45]. The relationship between hospitalization—related loss of muscle mass and strength and declining functional capacity and the risk for future hospital admission needs further exploration.

Collectively, malnutrition, loss of muscle mass, strength, and functional capacity are accelerated in hospitalized older adults [55, 61, 62, 66]. Increased LOS worsens malnutrition and sarcopenia, creating a vicious cycle of disease severity, increased frequency and severity of complications, increasing LOS, and rehospitalization rates [62].

Due to their associated comorbid conditions, malnutrition and sarcopenia impose a major economic burden on healthcare systems, contributing to escalating healthcare costs. A study of hospitalized patients showed significantly increased costs in malnourished patients. The mean daily expense was $228.00 per malnourished patients versus $138.00 per well-nourished patients, a cost increase of 60.5%, after including costs for medications and tests, the cost to treat malnourished patients rose by 308.9% [53]. According to a British Association of Parenteral and Enteral Nutrition report, malnutrition in the UK costs in excess of £13 billion per year: £8 billion is for healthcare, including hospital inpatients and outpatients, and primary care (prescriptions and general medical services), and £5 billion is for nursing, residential, and home care services [67]. Additionally, sarcopenia significantly increases health care costs. A 2000 US study estimated that sarcopenia resulted in $18.5 billion dollars in direct health care expenditures, which reflected 1.5% of total healthcare expenditures. Moreover, a 10% reduction in sarcopenia prevalence would result in $1.1 billion US dollars in health care savings [68].

TABLE 2: The clinical signs and symptoms of Malnutrition-Sarcopenia Syndrome.

Malnutrition-Sarcopenia Syndrome	
Malnutrition	Sarcopenia
↓ Food intake	↓ Muscle mass
↓ Appetite	↓ Muscle strength and/or functionality
↓ Body weight	

2.4 SCREENING AND ASSESSING FOR MALNUTRITION SARCOPENIA SYNDROME

Clinicians should integrate nutrition assessment with sarcopenia screening for optimal evaluation of these two inter-related nutritional issues to help improve patients' clinical outcomes. A variety of malnutrition screening tools are available such as the Malnutrition Screening Tool (MST) [69], Malnutrition Universal Screening Tool (MUST) [70], the short form of the Mini-Nutritional Assessment (SF-MNA) [71] and Nutrition Risk Screening-2002 (NRS-2002) [72]. A standard for nutritional assessment, the Subjective Global Assessment (SGA), is a valid and reliable method to assess nutritional status in a variety of patient populations [73]. The MNA is another reliable assessment tool validated for use with older adults in multiple settings [71, 74]. Elements of history and physical examination are commonly shared among these tools and include unintentional weight loss (e.g., 3 kg within the last 3 months), decreased food intake, gastrointestinal symptoms, and functional impairment. For sarcopenia screening, a simple clinician tool has been suggested by the European Geriatric Medical Society (EUGMS) Consensus Committee of defining sarcopenia (Figure 1), in which older adults are screened and assessed for sarcopenia using both gait speed and handgrip strength measurements [9, 75–77]. If a patient is identified to have slow gait speed or low hand grip strength, muscle mass should then be measured and evaluated. Based on the evidence presented, the combination of screening and assessing for malnutrition and sarcopenia is recommended to screen for the presence of MSS in at-risk patient populations, particularly older adults in clinical settings and in the community. The proposed clinical signs and symptoms to identify MSS are highlighted in Table 2. Specifically, to facilitate screening and assessment of MSS, we propose that patients would be at high risk for MSS if at least four of these criteria are present:

1. recent history of reduced appetite that resulted in poor food intake,
2. unintentional weight loss of 3 kg or more over the last 3 months,
3. low muscle mass (as measured by DXA, CT, MRI, or BIA),
4. decreased gait speed (less than 0.8 meter/second), and
5. reduced hand grip strength for age and gender.

Future research is warranted to determine the reliability and validity of this assessment tool across patient populations and settings.

2.5 CONCLUSION

Malnutrition and sarcopenia are both commonly occurring conditions across patient populations, especially older adults. Both conditions result in numerous and substantial negative outcomes to both the patient and the health care system, including increased morbidity and mortality, decreased patient quality of life and functionality, and increased health care costs and rehospitalization rates. Historically, patients have been screened or assessed by healthcare practitioners for either malnutrition or sarcopenia, but rarely for both conditions concurrently. However, many patients present clinically with both conditions in parallel and this combination, or the malnutrition sarcopenia syndrome should be the focus of future nutrition screening and assessment in at-risk patient populations. Examining the entirety of the patient's nutritional and functional status through screening and assessment for both malnutrition and sarcopenia will enable healthcare practitioners to better determine the presence of MSS in their patients and target interventions to fit the patients' needs. Moreover, as the world is aging and older adults will utilize health care services at an increased rate, this could ultimately result in better patient care and outcomes in this unique and expanding patient population. Clinicians and researchers are called upon to work together to develop a practical, reliable, and valid tool for MSS that is appropriate for implementation into a variety of clinical practice settings, with the aim of identifying patients with MSS and providing the appropriately targeted interventions.

REFERENCES

1. M. Elia and Malnutrition Advisory Group, Guidelines for Detection and Management of Malnutrition, BAPEN, Maidenhead, UK, 2000.
2. R. J. Stratton, C. J. Green, and M. Elia, Disease-Related Malnutrition: An Evidence-Based Approach to Treatment, CABI, Wallingford, UK, 2003.
3. G. L. Jensen, J. Mirtallo, C. Compher et al., "Adult starvation and disease-related malnutrition: a proposal for etiology-based diagnosis in the clinical practice setting

from the International Consensus Guideline Committee," Clinical Nutrition, vol. 29, no. 2, pp. 151–153, 2010.

4. R. H. Demling, "Nutrition, anabolism, and the wound healing process: an overview," ePlasty, vol. 9, pp. 65–94, 2009.

5. R. R. Wolfe, "The underappreciated role of muscle in health and disease," American Journal of Clinical Nutrition, vol. 84, no. 3, pp. 475–482, 2006.

6. J. J. McCarthy and K. A. Esser, "Anabolic and catabolic pathways regulating skeletal muscle mass," Current Opinion in Clinical Nutrition and Metabolic Care, vol. 13, no. 3, pp. 230–235, 2010.

7. K. S. Nair, "Muscle protein turnover: methodological issues and the effect of aging," Journals of Gerontology A, vol. 50, pp. 107–112, 1995.

8. I. H. Rosenberg, "Sarcopenia: origins and clinical relevance," Journal of Nutrition, vol. 127, no. 5, supplement, pp. 990S–991S, 1997.

9. A. J. Cruz-Jentoft, J. P. Baeyens, J. M. Bauer et al., "Sarcopenia: European consensus on definition and diagnosis," Age and Ageing, vol. 39, no. 4, pp. 412–423, 2010.

10. K. Norman, N. Stobäus, M. C. Gonzalez, J. D. Schulzke, and M. Pirlich, "Hand grip strength: outcome predictor and marker of nutritional status," Clinical Nutrition, vol. 30, no. 2, pp. 135–142, 2011.

11. M. Pirlich, T. Schütz, K. Norman et al., "The German hospital malnutrition study," Clinical Nutrition, vol. 25, no. 4, pp. 563–572, 2006.

12. M. Pirlich, T. Schütz, M. Kemps et al., "Social risk factors for hospital malnutrition," Nutrition, vol. 21, no. 3, pp. 295–300, 2005.

13. S. Forster and S. Gariballa, "Age as a determinant of nutritional status: a cross sectional study," Nutrition Journal, vol. 4, article 28, 2005.

14. L. C. P. G. M. de Groot, A. M. Beck, M. Schroll, and W. A. van Staveren, "Evaluating the DETERMINE your nutritional health checklist and the mini nutritional assessment as tools to identify nutritional problems in elderly Europeans," European Journal of Clinical Nutrition, vol. 52, no. 12, pp. 877–883, 1998.

15. Y. Guigoz, "The Mini Nutritional Aseesment (MNA) review of the literature—what does it tell us?" The Journal of Nutrition, Health & Aging, vol. 10, no. 6, pp. 466–487, 2006.

16. D. Paddon-Jones, Lean Body Mass Loss With Age, Abbott Nutrition, Columbus, Ohio, USA, 2009, Edited by: J. Gussler.

17. A. B. Newman, J. S. Lee, M. Visser et al., "Weight change and the conservation of lean mass in old age: The Health, Aging and Body Composition Study," American Journal of Clinical Nutrition, vol. 82, no. 4, pp. 872–878, 2005.

18. G. B. Forbes, "Longitudinal changes in adult fat-free mass: influence of body weight," American Journal of Clinical Nutrition, vol. 70, no. 6, pp. 1025–1031, 1999.

19. M. Karlsson, J. Å. Nilsson, I. Sernbo, I. Redlund-Johnell, O. Johnell, and K. J. Obrant, "Changes of bone mineral mass and soft tissue composition after hip fracture," Bone, vol. 18, no. 1, pp. 19–22, 1996.

20. F. Landi, R. Liperoti, A. Russo, et al., "Sarcopenia as a risk factor for falls in elderly individuals: results from the ilSIRENTE study," Clinical Nutrition. In press.

21. F. Landi, R. Liperoti, D. Fusco, et al., "Prevalence and risk factors of sarcopenia among nursing home older residents," Journals of Gerontology A, vol. 67, pp. 48–55, 2012.

22. I. Janssen, S. B. Heymsfield, and R. Ross, "Low relative skeletal muscle mass (sarcopenia) in older persons is associated with functional impairment and physical disability," Journal of the American Geriatrics Society, vol. 50, no. 5, pp. 889–896, 2002.

23. T. Cederholm, C. Jagren, and K. Hellstrom, "Outcome of protein-energy malnutrition in elderly medical patients," American Journal of Medicine, vol. 98, no. 1, pp. 67–74, 1995.

24. A. B. Newman, D. Yanez, T. Harris, A. Duxbury, P. L. Enright, and L. P. Fried, "Weight change in old age and its association with mortality," Journal of the American Geriatrics Society, vol. 49, no. 10, pp. 1309–1318, 2001.

25. J. I. Wallace, R. S. Schwartz, A. Z. LaCroix, R. F. Uhlmann, and R. A. Pearlman, "Involuntary weight loss in older outpatients: incidence and clinical significance," Journal of the American Geriatrics Society, vol. 43, no. 4, pp. 329–337, 1995.

26. D. F. Williamson and E. R. Pamuk, "The association between weight loss and increased longevity: a review of the evidence," Annals of Internal Medicine, vol. 119, no. 7, pp. 731–736, 1993.

27. S. A. French, A. R. Folsom, R. W. Jeffery, and D. F. Williamson, "Prospective study of intentionality of weight loss and mortality in older women: The Iowa Women's Health Study," American Journal of Epidemiology, vol. 149, no. 6, pp. 504–514, 1999.

28. J. L. Locher, D. L. Roth, C. S. Ritchie et al., "Body mass index, weight loss, and mortality in community-dwelling older adults," Journals of Gerontology A, vol. 62, no. 12, pp. 1389–1392, 2007.

29. E. Cereda, C. Pedrolli, A. Zagami et al., "Body mass index and mortality in institutionalized elderly," Journal of the American Medical Directors Association, vol. 12, no. 3, pp. 174–178, 2011.

30. S. S. Han, K. W. Kim, K. I. Kim et al., "Lean mass index: a better predictor of mortality than body mass index in elderly Asians," Journal of the American Geriatrics Society, vol. 58, no. 2, pp. 312–317, 2010.

31. F. Landi, R. Liperoti, D. Fusco, et al., "Sarcopenia and mortality among older nursing home residents," Journal of the American Medical Directors Association, vol. 13, no. 2, pp. 121–126, 2011.

32. D. Bunout, M. P. de la Maza, G. Barrera, L. Leiva, and S. Hirsch, "Association between sarcopenia and mortality in healthy older people," Australasian Journal on Ageing, vol. 30, no. 2, pp. 89–92, 2011.

33. R. Cooper, D. Kuh, and R. Hardy, "Objectively measured physical capability levels and mortality: systematic review and meta-analysis," British Medical Journal, vol. 341, Article ID c4467, 2010.

34. E. M. Castillo, D. Goodman-Gruen, D. Kritz-Silverstein, D. J. Morton, D. L. Wingard, and E. Barrett-Connor, "Sarcopenia in elderly men and women: The Rancho Bernardo Study," American Journal of Preventive Medicine, vol. 25, no. 3, pp. 226–231, 2003.

35. I. Janssen, "Influence of sarcopenia on the development of physical disability: The Cardiovascular Health Study," Journal of the American Geriatrics Society, vol. 54, no. 1, pp. 56–62, 2006.

36. S. M. Schneider, P. Veyres, X. Pivot et al., "Malnutrition is an independent factor associated with nosocomial infections," British Journal of Nutrition, vol. 92, no. 1, pp. 105–111, 2004.

37. J. Edington, J. Boorman, E. R. Durrant et al., "Prevalence of malnutrition on admission to four hospitals in England," Clinical Nutrition, vol. 19, no. 3, pp. 191–195, 2000.

38. G. Cosquëric, A. Sebag, C. Ducolombier, C. Thomas, F. Piette, and S. Weill-Engerer, "Sarcopenia is predictive of nosocomial infection in care of the elderly," British Journal of Nutrition, vol. 96, no. 5, pp. 895–901, 2006.

39. A. Vivanti, N. Ward, and T. Haines, "Nutritional status and associations with falls, balance, mobility and functionality during hospital admission," The Journal of Nutrition, Health & Aging, vol. 15, no. 5, pp. 388–391, 2011.

40. S. A. Neumann, M. D. Miller, L. Daniels, and M. Crotty, "Nutritional status and clinical outcomes of older patients in rehabilitation," Journal of Human Nutrition and Dietetics, vol. 18, no. 2, pp. 129–136, 2005.

41. J. Potter, K. Klipstein, J. J. Reilly, and M. Roberts, "The nutritional status and clinical course of acute admissions to a geriatric unit," Age and Ageing, vol. 24, no. 2, pp. 131–136, 1995.

42. R. Muhlethaler, A. E. Stuck, C. E. Minder, and B. M. Frey, "The prognostic significance of protein-energy malnutrition in geriatric patients," Age and Ageing, vol. 24, no. 3, pp. 193–197, 1995.

43. M. Lumbers, L. T. Driver, R. J. Howland, M. W. J. Older, and C. M. Williams, "Nutritional status and clinical outcome in elderly female surgical orthopaedic patients," Clinical Nutrition, vol. 15, no. 3, pp. 101–107, 1996.

44. T. M. Gill, H. G. Allore, T. R. Holford, and Z. Guo, "Hospitalization, restricted activity, and the development of disability among older persons," Journal of the American Medical Association, vol. 292, no. 17, pp. 2115–2124, 2004.

45. M. M. Suesada, M. A. Martins, and C. R. Carvalho, "Effect of short-term hospitalization on functional capacity in patients not restricted to bed," American Journal of Physical Medicine and Rehabilitation, vol. 86, no. 6, pp. 455–462, 2007.

46. T. Lang, T. Streeper, P. Cawthon, K. Baldwin, D. R. Taaffe, and T. B. Harris, "Sarcopenia: etiology, clinical consequences, intervention, and assessment," Osteoporosis International, vol. 21, no. 4, pp. 543–559, 2010.

47. M. Visser, B. H. Goodpaster, S. B. Kritchevsky et al., "Muscle mass, muscle strength, and muscle fat infiltration as predictors of incident mobility limitations in well-functioning older persons," Journals of Gerontology A, vol. 60, no. 3, pp. 324–333, 2005.

48. W. J. Evans, "Effects of exercise on body composition and functional capacity of the elderly," Journals of Gerontology A, vol. 50, pp. 147–150, 1995.

49. K. F. Reid, E. N. Naumova, R. J. Carabello, E. M. Phillips, and R. A. Fielding, "Lower extremity muscle mass predicts functional performance in mobility-limited elders," The Journal of Nutrition, Health & Aging, vol. 12, no. 7, pp. 493–498, 2008.

50. W. R. Frontera, D. Suh, L. S. Krivickas, V. A. Hughes, R. Goldstein, and R. Roubenoff, "Skeletal muscle fiber quality in older men and women," American Journal of Physiology, vol. 279, no. 3, pp. C611–C618, 2000.

51. J. E. Morley, A. M. Abbatecola, J. M. Argiles, et al., "Sarcopenia with limited mobility: an international consensus," Journal of the American Medical Directors Association, vol. 12, pp. 403–409, 2011.

52. J. Dzieniszewski, M. Jarosz, B. Szczygieł et al., "Nutritional status of patients hospitalised in Poland," European Journal of Clinical Nutrition, vol. 59, no. 4, pp. 552–560, 2005.

53. M. I. Correia and D. L. Waitzberg, "The impact of malnutrition on morbidity, mortality, length of hospital stay and costs evaluated through a multivariate model analysis," Clinical Nutrition, vol. 22, no. 3, pp. 235–239, 2003.

54. M. Suominen, S. Muurinen, P. Routasalo et al., "Malnutrition and associated factors among aged residents in all nursing homes in Helsinki," European Journal of Clinical Nutrition, vol. 59, no. 4, pp. 578–583, 2005.

55. V. A. Leandro-Merhi, J. L. de Aquino, and J. F. Sales Chagas, "Nutrition status and risk factors associated with length of hospital stay for surgical patients," Journal of Parenteral and Enteral Nutrition, vol. 35, no. 2, pp. 241–248, 2011.

56. C. S. Chima, K. Barco, M. L. Dewitt, M. Maeda, J. C. Teran, and K. D. Mullen, "Relationship of nutritional status to length of stay, hospital costs, and discharge status of patients hospitalized in the medicine service," Journal of the American Dietetic Association, vol. 97, no. 9, pp. 975–978, 1997.

57. D. H. Sullivan and R. C. Walls, "Protein-energy undernutrition and the risk of mortality within six years of hospital discharge," Journal of the American College of Nutrition, vol. 17, no. 6, pp. 571–578, 1998.

58. N. Allaudeen, A. Vidyarthi, J. Maselli, and A. Auerbach, "Redefining readmission risk factors for general medicine patients," Journal of Hospital Medicine, vol. 6, no. 2, pp. 54–60, 2011.

59. A. M. Mudge, K. Kasper, A. Clair et al., "Recurrent readmissions in medical patients: a prospective study," Journal of Hospital Medicine, vol. 6, no. 2, pp. 61–67, 2011.

60. J. Edington, J. Boorman, E. R. Durrant et al., "Prevalence of malnutrition on admission to four hospitals in England," Clinical Nutrition, vol. 19, no. 3, pp. 191–195, 2000.

61. P. M. Cawthon, K. M. Fox, S. R. Gandra et al., "Do muscle mass, muscle density, strength, and physical function similarly influence risk of hospitalization in older adults?" Journal of the American Geriatrics Society, vol. 57, no. 8, pp. 1411–1419, 2009.

62. D. E. Alley, A. Koster, D. MacKey et al., "Hospitalization and change in body composition and strength in a population-based cohort of older persons," Journal of the American Geriatrics Society, vol. 58, no. 11, pp. 2085–2091, 2010.

63. L. Ferrucci, J. M. Guralnik, M. Pahor, M. C. Corti, and R. J. Havlik, "Hospital diagnoses, medicare charges, and nursing home admissions in the year when older persons become severely disabled," Journal of the American Medical Association, vol. 277, no. 9, pp. 728–734, 1997.

64. M. A. Sager, T. Franke, S. K. Inouye et al., "Functional outcomes of acute medical illness and hospitalization in older persons," Archives of Internal Medicine, vol. 156, no. 6, pp. 645–652, 1996.

65. C. Pichard, U. G. Kyle, A. Morabia, A. Perrier, B. Vermeulen, and P. Unger, "Nutritional assessment: lean body mass depletion at hospital admission is associated with an increased length of stay," American Journal of Clinical Nutrition, vol. 79, no. 4, pp. 613–618, 2004.

66. P. Kortebein, A. Ferrando, J. Lombeida, R. Wolfe, and W. J. Evans, "Effect of 10 days of bed rest on skeletal muscle in healthy older adults," Journal of the American Medical Association, vol. 297, no. 16, pp. 1772–1774, 2007.

67. M. Elia and C. A. Russell, "Combating malnutrition: recommendations for action. Report From the Advisory Group on Malnutrition," Led By BAPEN; Output of a meeting of the Advisory Group on Malnutrition, 12 June 2008.

68. I. Janssen, D. S. Shepard, P. T. Katzmarzyk, and R. Roubenoff, "The healthcare costs of sarcopenia in the United States," Journal of the American Geriatrics Society, vol. 52, no. 1, pp. 80–85, 2004.

69. M. Ferguson, S. Capra, J. Bauer, and M. Banks, "Development of a valid and reliable malnutrition screening tool for adult acute hospital patients," Nutrition, vol. 15, no. 6, pp. 458–464, 1999.

70. M. Elia, Screening for Malnutrition: A Multidisciplinary Responsibility. Development and Use of the 'Malnutrition Universal Screening Tool' ('MUST') for Adults, BAPEN, Worcestershire, UK, 2003.

71. Y. Guigoz, S. Lauque, and B. J. Vellas, "Identifying the elderly at risk for malnutrition the mini nutritional assessment," Clinics in Geriatric Medicine, vol. 18, no. 4, pp. 737–757, 2002.

72. J. Kondrup, H. H. Ramussen, O. Hamberg et al., "Nutritional risk screening (NRS 2002): a new method based on an analysis of controlled clinical trials," Clinical Nutrition, vol. 22, no. 3, pp. 321–336, 2003.

73. H. M. Kruizenga, M. W. Van Tulder, J. C. Seidell, A. Thijs, H. J. Ader, and M. A. van Bokhorst-de van der Schueren, "Effectiveness and cost-effectiveness of early screening and treatment of malnourished patients," American Journal of Clinical Nutrition, vol. 82, no. 5, pp. 1082–1089, 2005.

74. M. J. Kaiser, J. M. Bauer, C. Rämsch et al., "Frequency of malnutrition in older adults: a multinational perspective using the mini nutritional assessment," Journal of the American Geriatrics Society, vol. 58, no. 9, pp. 1734–1738, 2010.

75. D. M. Buchner, E. B. Larson, E. H. Wagner, T. D. Koepsell, and B. J. de Lateur, "Evidence for a non-linear relationship between leg strength and gait speed," Age and Ageing, vol. 25, no. 5, pp. 386–391, 1996.

76. H. C. Roberts, H. J. Denison, H. J. Martin et al., "A review of the measurement of grip strength in clinical and epidemiological studies: towards a standardised approach," Age and Ageing, vol. 40, no. 4, pp. 423–429, 2011.

77. A. J. Cruz-Jentoft, F. Landi, E. Topinková, and J. P. Michel, "Understanding sarcopenia as a geriatric syndrome," Current Opinion in Clinical Nutrition and Metabolic Care, vol. 13, no. 1, pp. 1–7, 2010.

CHAPTER 3

Novel Insights on Nutrient Management of Sarcopenia in Elderly

MARIANGELA RONDANELLI, MILENA FALIVA,
FRANCESCA MONTEFERRARIO, GABRIELLA PERONI,
ERICA REPACI, FRANCESCA ALLIERI, AND SIMONE PERNA

3.1 INTRODUCTION

Sarcopenia is defined by the European Working Group on Sarcopenia in Older People (EWGSOP) [1] as a syndrome characterized by progressive and generalized loss of muscle mass and strength. Sarcopenia is a physiological phenomenon that usually starts in the fifth decade. van Kan (2009) has investigated the prevalence of sarcopenia in the population aged 60–70 years: in this age group, the prevalence ranged from 5 to 13% but increased to 11–50% in subjects aged >80 years [2].

Sarcopenia becomes responsible not only for the reduction of mobility and the level of autonomy of the elderly, but also for their ability to maintain good health. The functional reduction of the quadriceps muscle predisposes to a limitation in walking, with risk of falls and fractures of the femoral neck. A survey conducted in the USA has estimated the cost-related health consequences of sarcopenia to be 20–30 billion dollars [3].

In most elderly patients, the onset of sarcopenia is multifactorial. Like in all body tissues, muscle proteins are subjected to a constant process of synthesis and degradation; in healthy adults (with an adequate protein intake) this turnover is in balance, allowing the maintenance of a positive nitrogen balance and a constant muscle mass [4, 5].

In elderly, one of the pathogenic mechanisms leading to sarcopenia is altered muscle protein metabolism: the proteolytic processes are not accompanied by an adequate protein synthesis within the physiological turnover, and muscle cells lose progressively the sensitivity to the ana-bolic stimulus induced from the essential leucine and IGF-1 (insulin-like growth factor), thus manifesting the so-called "anabolic resistance" [6].

This phenomenon may be associated with other hormonal, functional, and nutritional factors, each of which may contribute to a greater or lesser extent—depending on gender, age, and clinical condition of the patient—to the progression of disease, defined as secondary sarcopenia [7]. With respect to nutritional causes, about 40% of subjects >70 years do not as-sume the current RDA (Recommended Dietary Allowances) of proteins (0.8 g/kg/day) [8]. The phenomenon depends upon several factors, each with a variable contribution: these include odontostomatological prob-lems, capable of altering the masticatory function (therefore influencing the choice of foods with reduced content of proteins); a reduced capacity of digestion and assimilation of proteins in enteric tube; delayed gastric emptying, associated with a reduced gallbladder contractility and higher serum levels of the hormone cholecystokinin (CCK) and neuropeptide Y (PYY) (facilitating a long-lasting satiety [9]); a higher blood concentra-tion of leptin in the elderly (showing that the anorexigenic signal prevails over the orexigenic one [10]); a propensity, increasing with age, to take sweet foods, easily chewable and already ready to eat, but not always ad-equate in the amino acid content.

In addition, costs are an issue, since a greater adherence to the Medi-terranean diet is inversely related to BMI but leads to higher cost, which in 2006 were estimated at 1.2 €/day [11], due to the higher cost of meat and fish, compared with carbohydrates. Lastly, it is necessary to add that, according to available evidence, current RDA that define the protein intake in the elderly population should be revised, because for a number of reasons—many of which have already been discussed—they are often inadequate in terms of quantity and quality [12–15].

The FAO and the WHO indicate that an intake of 0.75 grams of high quality protein per kilogram of body weight is safe and adequate; however, for elderly subjects, it has been proposed to increase this value to 1.25 g/kg/day in order to avoid sarcopenia [16]. It is also necessary to consider that elderly subjects frequently present subclinical nutritional deficits, in particular of vitamins and minerals useful for the muscular tropism, such as vitamin D [17].

TABLE 1: Nutrients and drugs that have been shown to present an activity of stimulation in increasing the mass and/or muscle strength in humans or in the animal model.

Nutrients	Proteins and amino acids (BCAAs) and creatine.
	Antioxidants (vitamin E, vitamin C, carotenoids, and resveratrol)
	Vitamins: vitamin D
	Long-chain omega-3 fatty acids
Drugs	Antagonists of mineral corticoids (Spironolactone)
	ACE inhibitors
Hormone replacement therapy	Testosterone (T)
	Growth hormone (GH)
	Combination therapy: T and GH
	Estrogen
	DHEA-S

The more rational approach to delay the progression of sarcopenia is based on the combination of proper nutrition, possibly associated with the use of supplements and/or foods for special medical purposes, and a regular exercise program. Alternative treatments which are based on administration of hormone preparations such as testosterone, GH, and estrogens are still not universally accepted and require further investigation [18, 19]. Table 1 lists the nutrients and drugs which have been shown to increase the mass and/or muscle strength in humans or animal models.

Table 2 summarizes the studies (prospective cohort studies or randomized controlled trials) performed in elderly subjects to investigate the optimum dietary supplementation, other than proteins/aminoacids, for the treatment of sarcopenia.

Given this background, the aim of the present narrative review is to summarize the state of the art according to the extant literature about two topics: (1) the correct intake of protein and amino acids, in particular branched chain amino acids (BCAAs) need for prevention and treatment of sarcopenia; (2) the correct intake of other nutrients, such as antioxidant, vitamin D, and long-chain omega-3 polyunsaturated fatty acids, or dietary supplements, such as beta-hydroxy-methylbutyrate and creatine, in need for prevention and treatment of sarcopenia.

3.2 METHODS

The present narrative review was performed following the steps by Egger et al. [20]. Table 3 showed the summary of methodology used. The step were (1) configuration of a working group: three operators skilled in endocrinology and clinical nutrition, of whom one acting as a methodological operator and two participating as clinical operators; (2) formulation of the revision question on the basis of considerations made in the abstract: "the state of the art on metabolic and nutritional correlates of sarcopenia and their nutritional treatment"; (3) identification of relevant studies: a research strategy was planned, on PUBMED (Public MedLine run by the National Center of Biotechnology Information (NCBI) of the National Library of Medicine of Bethesda (USA)), as shown in Table 3; (4) analysis and presentation of the outcomes: the data extrapolated from the revised studies were carried out in the form of a narrative review of the reports and were collocated in tables. The flow diagram of narrative review of the literature has been reported in Figure 1. At the beginning of each section, the keywords considered and the kind of studies chosen have been reported. Suitable for the narrative review were prospective cohort studies, randomized controlled trials (RCT), reviews, meta-analyses, cross sectional studies, and position paper which considered elderly with diagnosis of sarcopenia defined by the European Working Group on Sarcopenia in Older People (EWGSOP) [1].

TABLE 2: Studies (prospective cohort study or randomized controlled trial) performed in elderly subjects to investigate the optimum dietary supplementation, other than proteins, for the treatment of sarcopenia.

Nutrients	Author	Type of study	Results	Recommended treatment
Vitamin D	Snijder et al., 2006 [114]	Prospective cohort study	Poor vitamin D status is independently associated with an increased risk of falling in the elderly, particularly in those aged 65–75 yr.	
	Verhaar et al., 2000 [109]	Randomized controlled trial	Six months of alphacalcidol treatment led to a significant increase in the walking distance over 2 minutes.	Six months of vitamin D treatment (0.5 microg alphacalcidol)
	Gloth et al., 1995 [110]	Randomized controlled trial	In this cohort of homebound older people, improvement in vitamin D status was associated with functional improvement as measured by the Frail Elderly Functional Assessment questionnaire.	One month of therapy with either placebo or vitamin D (ergo-calciferol)
Beta-hydroxy-beta-methylbutyrate (HMB)	Flakoll et al., 2004 [50]	Randomized controlled trial	Daily supplementation of HMB, arginine, and lysine for 12 wk	Daily supplementation of HMB, arginine, and lysine for 12 wk positively altered measurements of functionality, strength, fat-free mass, and protein synthesis, suggesting that the strategy of targeted nutrition has the ability to affect muscle health in elderly women.
Long-chain omega-3 fatty acids	Smith et al., 2011 [131]	Randomized controlled trial	Omega-3 fatty acid supplementation had no effect on the basal rate of muscle protein synthesis but enhanced the hyperaminoacidemia-hyperinsulinemia-induced increase in the rate of muscle protein synthesis, which was accompanied by greater increases in muscle mTORSer2448 phosphorylation	1.86 g eicosapentaenoic acid (EPA, 20:5n23) and 1.50 g docosahexaenoic acid (DHA, 22:6n23), both as ethyl esters

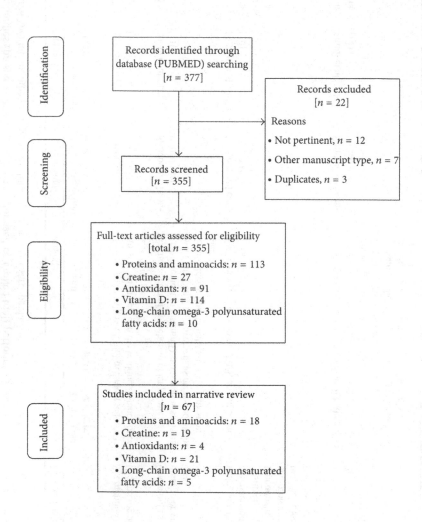

FIGURE 1: Flow diagram of narrative review of literature.

TABLE 3: Summary of methodology.

Step	General activities	Specific activities
Step 1	Configuration of a working group	Three operators skilled in clinical nutrition:
		(i) one operator acting as a methodological operator
		(ii) two operators participating as clinical operators
Step 2	Formulation of the revision question	Evaluation of the state of the art on metabolic and nutritional correlates of sarcopenia and their nutritional treatment
Step 3	Identification of relevant studies on PUBMED	(a) Definition of the key words (sarcopenia, nutrients, and dietary supplement), allowing the definition of the interest field of the documents to be searched, grouped in inverted commas ("…"), and used separately or in combination;
		(b) use of the Boolean (a data type with only two possible values: true or false) AND operator that allows the establishments of logical relations among concepts;
		(c) research modalities: advanced search;
		(d) limits: time limits: papers published in the last 20 years; humans; languages: English;
		(e) manual search performed by the senior researchers experienced in clinical nutrition through the revision of reviews and individual articles on sarcopenia in elderly published in journals qualified in the Index Medicus
Step 4	Analysis and presentation of the outcomes	The data extrapolated from the revised studies were carried out in the form of a narrative review of the reports and were collocated in tables.

3.3 RESULTS

3.3.1 AMINOACIDS AND PROTEIN

This research has been carried out based on the keywords: "sarcopenia" AND "proteins" AND "aminoacids"; 113 articles were sourced. Among them, 1 observation study, 6 reviews, 2 cross sectional studies, 8 randomized controlled studies (RCT), and 1 position paper have been selected and discussed.

It is known that the amino acids, including branched chain amino acids (BCAAs), are necessary for the maintenance of muscle health in the elderly [21]. Approximately 300–600 grams of muscle proteins is degraded and resynthesized daily, with complete renewal of the pool of muscle protein in the human body occurring every 3-4 months. Food intake stimulates muscle protein synthesis, resulting in a positive protein balance. After taking a protein-rich meal, the degree of protein synthesis remains elevated for more than 5 hours, with a peak of 2-3 hours after the intake [22]. It has been shown that in adult subjects a dose of approximately 15–20 grams of protein (or 7.5 grams of essential amino acids) is sufficient to maximize the synthesis of muscle proteins [23].

Probably the elderly, compared with younger subjects, would require a larger amount of protein to obtain the same maximization of protein synthesis, probably 30 grams as shown by Pennings et al. [24]. The bioavailability of amino acids plays a major role in the regulation of protein metabolism in elderly subjects and therefore a nutritional therapy must necessarily aim also at the recovery of muscle and sensitivity to the stimulus induced by the protein-synthetic amino acids, in order to contrast anabolic resistance. Over the past few years, the analysis of the different nutritional strategies has allowed the definition of some key concepts, recently discussed in a position paper by the PROT-AGE Study group [16], which include the recommended amount of protein intake for the healthy elderly; the recommended amount of protein intake for the elderly with acute or chronic conditions; the role of physical activity in association with dietary intake to maintain muscle strength and function in elderly; the practical administration of food proteins (source and quality of dietary protein, protein intake timing, and energy intake).

3.3.1.1 RECOMMENDED AMOUNT OF PROTEIN INTAKE FOR HEALTHY ELDERLY

To maintain and recover the muscle, elderly subjects need to have a greater protein intake compared with younger subjects; older people should have an average intake of protein of 1.2/g/kg/day of body weight/day [16]. The threshold for anabolic meal intake of protein/amino acids must be greater in elderly subjects (i.e., 25 to 30 g of protein per meal, containing approximately 2.5–2.8 g of leucine), compared with young adults [24, 25]. The source of protein, timing of intake, and supplementation with amino acid supplements should be considered when making recommendations on the intake of dietary protein in the elderly.

3.3.1.2 RECOMMENDATIONS IN PROTEIN INTAKE IN THE COURSE OF ACUTE AND CHRONIC PATHOLOGY

In subjects with pathological conditions, the amount of additional proteins to be taken or the protein requirement depends on the specific disease, its severity, nutritional status of the subject before the onset of the disease, and the impact of the disease on the state of nutrition. The majority of elderly patients who present an acute or chronic disease have an increased need for protein intake (1.2 to 1.5 g/kg body weight/day), while patients with critical illnesses or severe malnutrition have a need of protein equal to 2 g/kg body weight/day. Elderly subjects with severe renal impairment (estimated glomerular filtration rate $< 30\,mL/min/1.73\,m^2$) who are not on dialysis are an exception and, conversely, must limit their protein intake.

3.3.1.3 QUALITY OF PROTEIN AND SPECIFIC AMINO ACIDS

Not all dietary proteins have the same kinetic properties: the rate of absorption of dietary amino acids and their effect on the regulation of protein metabolism are dependent on the molecular characteristics of the protein. This characteristic gave rise to the distinction of dietary protein between fast and slow [26–28].

Previous work suggests that whey protein ingestion results in greater postprandial protein retention than does casein ingestion [29, 30].

The greater anabolic properties of whey than of casein are mainly attributed to the faster digestion and absorption kinetics of whey, which results in a greater increase in postprandial plasma amino acid availability and thereby further stimulates muscle protein synthesis [23, 26, 27, 31, 32].

Besides differences in protein digestion and absorption kinetics, whey and casein also markedly differ in their amino acid composition [26, 27, 32].

Whereas both proteins contain all the amino acids required to effectively stimulate muscle protein synthesis [33], whey has a considerably higher leucine content.

As regards differences between animal and vegetal sources, even if previous studies demonstrated that consumption of a meat-containing diet contributed to greater gains in fat-free mass and skeletal muscle mass with resistance training in older men than did a lactoovovegetarian diet [34, 35], more recent studies suggested that increases in muscle strength and size were not influenced by the predominant source of protein consumed by older men with adequate total protein intake [36].

3.3.1.4 BRANCHED CHAIN AMINO ACIDS AND LEUCINE

It has been suggested that leucine, which is an essential amino acid belonging to the category of the branched chain amino acids (BCAAs; valine, and together with the isoleucine, whose average requirement is 40 mg/kg/day), is critical to maintaining a healthy muscle tissue and liver. The main sources of leucine are chicken and fish, cottage cheese, lentils, sesame, and peanuts. Unlike many other amino acids, BCAAs are metabolized only in skeletal muscle, since the BCAA amino-transferase enzyme is not present in the liver, the site in which the enzymes metabolizing all other amino acids are present in maximum concentrations, and up to 58% of all the amino acids ingested (except BCAA) can be oxidized in the liver on the first pass. Skeletal muscle is able to oxidize only 6 amino acids during exercise: in addition to BCAAs, asparagine, aspartate, and glutamate. When combined with exercise training, BCAA supplementation increases testosterone and decreases cortisol to create an anabolic environment [37]. BCAAs represent 14%–18% of the total amino acid content of skeletal muscles [38]. At rest, BCAAs, in particular leucine, have an anabolic effect by increasing protein synthesis and/or a reducing the rate of protein degradation, resulting in a positive net muscle protein balance [39]. The infusion of BCAAs in humans elevates the phosphorylation and the activation of p70S6 kinase and 4E-BP1 in skeletal muscle [40]. Both p70S6 kinase and 4E-BP1 are downstream components of the mTOR signaling pathway, which controls RNA translation and synthesis of proteins, and which is recognized as the central node to support muscle hypertrophy [41]. Leucine is involved in the direct phosphorylation and activation of mTOR in skeletal muscle, further enhancing the protein synthetic response [42]. According to the WHO, for healthy people the daily demand of BCAA to cope with the normal loss in protein metabolism and turnover are the following: valine: 10 mg/kg body weight; isoleucine: 10 mg/kg body weight; leucine: 10 mg/kg body weight [43].

It was recognized that the leucine content of the meal is an important regulator of the synthesis of muscle proteins and influences body composition in the long term [44]. Further research has compared the intake of 10 grams of protein with 18% of leucine with a similar beverage containing 35% of leucine, concluding that the beverage with the highest concentration of leucine determines a greater signaling of protein synthesis, resulting in an inferior muscle catabolism by cortisol [45]. Other studies on leucine show that once the minimum requirement of leucine for protein synthesis is satisfied, leucine can be used to activate various signaling pathways, including mTOR. mTOR is a major regulator of protein synthesis, energy sensors, and sensors of nutrients and the availability of amino acids, particularly leucine. The mTOR pathway is activated when ATP levels are high and is blocked when ATP levels are reduced. The activation of mTOR is vital for skeletal muscle hypertrophy. Leucine presents significant activity to stimulate insulin synthesis, which may increase the availability of amino acids for the synthesis of muscle proteins; in addition, leucine inhibits the destruction of muscle proteins with consequent increased balance over time [46]. Although leucine is described as the most important of the three BCAAs, isoleucine and valine also play a role, although they have not shown the same potential as leucine. In fact, the hypertrophy induced by leucine decreases to zero as soon as the presence of the other two BCAAs is poor; regardless of the amount of leucine available to the muscles, muscle growth does not occur if the concentration of the other two BCAAs decreases below a given level [47].

3.3.1.5 BETA-HYDROXY-BETA-METHYLBUTYRATE

Beta-hydroxy-beta-methylbutyrate (HMB) is a product of leucine metabolism that has been shown to slow protein breakdown in muscle tissue [48]. HMB may be effective at limiting the demands placed on the elderly subjects by acute stresses, such as sudden increases in physical activity, an immunologic challenge, or acute malnutrition [48, 49].

Daily supplementation of HMB (2 g/day), arginine, and lysine for 12 wk positively altered measurements of functionality, strength, fat-free mass, and protein synthesis, suggesting that the strategy of targeted nutrition has the ability to affect muscle health in elderly women [50].

In conclusion, an adequate intake of proteins (1.2/g/kg/day) is essential to prevent sarcopenia and aminoacids supplementation; in particular branched chain amino acids (leucine 2.5 g/day) as well as the intake of beta-hydroxy butyrate (2 g/day) is a well documented intervention for treating sarcopenia.

3.3.2 CREATINE

This research has been carried out based on the keywords: "sarcopenia" AND "creatine"; 27 articles were sourced. Among them, 3 reviews, 11 RCT, 1 single blind study, 1 control case study, 2 observational studies, and 1 cross over study have been selected and discussed.

Creatine is chemically known as a nonprotein nitrogenous compounds; it is a tripeptide composed of three amino acids (glycine, arginine, and methionine). In the human body, creatine is synthesized in the liver and pancreas from the amino acids arginine, glycine, and methionine. Moreover, creatine is present in foods (meat and fish) and is taken with the diet in the amount of 1-2 grams per day. Approximately 95% of the creatine in the body is stored in skeletal muscles, as phosphocreatine (PCr) for about two-thirds of the total content, while the remaining part is stored as free creatine. The energy provided for the phosphorylation of adenosine diphosphate (ADP) to adenosine triphosphate (ATP) during and after intense exercise largely depends on the amount of PCr stored in the muscle.

With the depletion of PCr during intense exercise, the availability of energy decreases due to the inability to resynthesize ATP in the amount required to keep the high-intensity exercise [51]. Age-associated reductions of creatine/phosphocreatine in skeletal muscle have been reported in some studies [52, 53], although not all studies agree [54, 55]. The reduction of muscle creatine is biologically plausible, due to aging and, possibly, to certain comorbidities, such as sarcopenia, and/or changes with age in behavior (reduced physical activity and/or changes in dietary behaviors, such as decreased intake of meat for edentulous). The type II muscle fibers have a higher content of phosphocreatine compared with type I fibers (86 against and 74 mmol/kg dm) [56], and sarcopenia is characterized by a preferential atrophy of type II fibers [57]. The progressive atrophy of type II fibers may therefore partly explain the reduced muscle creatine in the elderly. Moreover, the reduction of creatine in the muscle of the elderly is in line with previous evidence that documents an increased oxidative process in aged skeletal muscles, for example, with decreased dependence on glycolysis [58] and a decrease of lactate dehydrogenase [59]. Smith et al. (1998) first reported an increase in muscle PCr (30%) in middle-aged adults (58 years) as a result of short-term intake of high doses of creatine (0.3 g/kg/day for 5 days) [52]. In a similar study, Rawson et al. (2002) reported a smaller increase in muscle phosphocreatine (7 versus 35%) in older (70 years) compared to younger subjects (24 years), in response to the ingestion of creatine (20 g/day for 5 days) [55]. However, the muscle PCr baseline was greater in the young compared with adult subjects described by Smith et al. (1998) [52], while the elderly subjects described by Rawson et al. (2002) had greater initial muscle PCr compared with younger subjects [55]. Brose et al. (2003) reported an increase in total muscle creatine (30% men, 17% women) in elderly subjects (70 years) who underwent 14 weeks of resistance training associated with intake of creatine in a dose of 5 g/day [60], a result that is similar to the increases reported in younger adults [61, 62]. Eijnde et al. (2003) reported an increase in total muscle creatine (5%) and free creatine (21%) following 6 months of exercise program for muscular endurance associated with the creatine supplementation (5 g/day) [63]. From these studies, although limited in number, it seems that the muscle creatine in the elderly can be increased with oral creatine supplementation in a dose of 5 g/day, but that the magnitude of the response can be significantly affected by the initial muscle creatine. Wyss et al. (1998) have suggested that the increase in extracellular creatine may decrease the absorption of creatine muscle by decreasing the activity of creatinine [64]. Although Rawson et al. (2002) have reported the presence of increased creatine in the blood of elderly subjects (elderly 68.5 mol/l, young 34.9 mol/l) [55], Tarnopolsky et al. (2003) showed no decrease in the activity of creatine after creatine ingestion in older men and women [65]. The most peculiar discovery was an improvement in fatigue resistance, which has been shown in many different studies using different exercise test [52, 66–70]. Some investigators have reported an increase in strength [68, 69], but this has not always been demonstrated [66, 67]. Importantly, in subsequent publications, researchers have begun to evaluate the performance of activities of daily living (activity daily living, ADL) and have shown that creatine supplementation may improve the performance of daily tasks identified in the ADL scale [69, 71, 72]. The improvement of the performance of activities of daily living is an important finding, because of the association between the performance of ADL, fall risk, and mortality. Among the studies that have evaluated the muscle mass, the majority showed a greater increase in lean mass accretion after ingesting creatine in combination with resistance training [60, 73, 74]. A further advantage given by combining creatine supplementation with resistance exercise is the increase in bone mineral content. Chilibeck et al. (2005) showed a greater increase (3.2 versus 1%) of bone mineral content in older men (71 years) after 12 weeks of creatine supplementation (0.3 g/kg for 5 days, 0.07 g/kg for 11 weeks), in combination with training against resistance compared to training alone [75]. Dalbo et al. (2009) have stated that creatine is an effective intervention to combat sarcopenia [76]. The timing of creatine ingestion (i.e., 0.03–0.5 g/kg before and after the sessions of resistance training) can be more relevant than the amount of creatine. These novel findings have immediate application for research and health professionals for the design of optimal creatine application strategies for older individuals [51].

In conclusion, an adequate creatine supplementation could represent an intriguing intervention to counteract sarcopenia, in particular fatigue related to sarcopenia, although double-blind, placebo-controlled studies have not been conducted.

3.3.3 ANTIOXIDANTS

This research has been carried out based on the keywords: "sarcopenia" AND "antioxidant"; 91 articles were sourced. Among them, 1 cross sectional study, 2 reviews, and 1 case control study have been selected and discussed.

Oxidative stress has been implicated as a central mechanism in the pathogenesis of sarcopenia [77].

Oxidative damage in skeletal muscle has been associated with the atrophy and loss of muscle function and fibers in sarcopenia [78].

Moreover, the accumulation of mitochondrial and nuclear DNA damage due to oxidative stress is thought to eventually compromise function, leading to the loss of myocytes [79].

Finally, reactive oxygen species can damage muscle tissue directly, but they also provide a trigger for the expression of inflammatory cytokines such as interleukin- (IL-) 1, tumor necrosis factor (TNF), and IL-6. In older age, a low-grade inflammatory state characterized by increased concentrations of inflammatory cytokines and acute phase proteins is common [80, 81].

Studies conducted among community-dwelling older adults suggest that the proinflammatory state does have a long-term consequence for sarcopenia. In the Longitudinal Aging Study Amsterdam, elevated IL-6 and CRP were associated with a loss of muscle strength over three years of follow-up [82].

Given this background, antioxidants (carotenoids, vitamin E, and vitamin C) should play an important role against sarcopenia.

Carotenoids inactivate free radicals and appear to modulate the transcription factors, such as the NFkB, which are involved in the regulation of IL-6 and other proinflammatory cytokines and have the ability, like alpha-tocopherol, to increase muscle strength [83, 84].

In the Women's Health and Aging Studies (WHAS) I and II, low serum carotenoid levels were associated with poor muscle strength [84]. Likewise in the InCHIANTI study, a low-carotene intake was associated with low physical performance [85]. These observations are consistent with a growing number of studies showing that a diet with high intake of fruits and vegetables is associated with a reduced risk of inflammation, hypertension, diabetes, cardiovascular disease, sarcopenia, and mortality [83].

Adherence to the Mediterranean diet, which is characterized by a high intake of fruits, vegetables, and whole grains, and lower consumption of red meat and saturated fats are associated with lower circulating IL-6 [86], and a recent trial showed the Mediterranean diet reduced IL-6 in adults [87].

In animal models, it was found that the ability of leucine to stimulate muscle protein synthesis is significantly decreased in aged rats compared to young adults. This defect was reversed when the animal was supplemented with antioxidants. The effects may be due to a reduction of the inflammatory state due to the antioxidants themselves [88]. In addition, the supplementation of vitamins E and C improves indices of oxidative stress associated with exercise in aged rats [89].

Concerning the specific vitamin E, in human studies, vitamin E has been shown to affect muscle strength of the elderly [90].

Several studies have shown the positive effects of vitamin E in reversing muscle damage during extensive muscle contraction (exercise) in healthy men. Vitamin E supplementation at a dose of 800 IU for 28 days resulted in lowering the expression of oxidative stress markers after a downhill run in both young and older men [91].

In another study, a longer supplementation period (12 weeks of vitamin E supplementation) lowered creatinine kinase level after exercise in young men, whereas older men showed decreased lipid peroxidation in both resting state and after exercise, indicating that vitamin E promotes adaptation against exercise induced-oxidative stress and reduced muscle damage [92].

In animal models, similar results were obtained [93, 94].

In conclusion, until today, despite the promising above-mentioned animal studies and studies on subjects without sarcopenia, no RCT studies have evaluated the efficacy of an integration with antioxidants in the elderly patient suffering from sarcopenia. These studies are needed, given that recent epidemiological studies in community-dwelling older adults show that low serum/plasma carotenoids are independently associated with low skeletal muscle strength and the development of walking disability.

3.3.4 VITAMIN D

This research has been carried out based on the keywords: "sarcopenia" AND "vitamin D"; 114 articles were sourced. Among them, 6 RCT, 5 reviews, 5 observational studies, 4 longitudinal studies, 2 population studies, 2 prospective studies, and 2 cohort analytic studies have been selected and discussed.

Vitamin D deficiency is common among geriatric patients (2–60%) [95, 96]. Vitamin D is hydroxylated in the liver to 25 (OH) D. This step is still well presented in the elderly, but it can be affected by liver disease [97, 98]. Further hydroxylation occurs in the kidney with the formation of 1,25-(OH2) D; however the activity of hydroxylation by the kidney may decrease with age, in parallel with the decline in renal function [99]. Consequences are the following: a low level of vitamin D, renal failure, and a low intake of calcium may result in mild secondary hyperparathyroidism. Increased levels of parathyroid hormone (PTH) cause an increase in bone turnover that is associated with bone loss, predominantly cortical; secondary hyperparathyroidism has been proposed as the main mechanism through which vitamin D deficiency contributes to the pathogenesis of hip fracture [95]. The presence of receptors for vitamin D was demonstrated in many organs [100], and the active metabolite, 1,25-2(OH)D, has been shown to be implicated in numerous systems that reduce the cell growth and inducing differentiation [101, 102]. Epidemiological studies have suggested that vitamin D deficiency is associated with cancer of the colon and breast [103, 104]. Furthermore, the status of vitamin D influences the immune system and insulin secretion [105].

Many studies have shown that low levels of 1,25-(OH) D and 25-(OH) D are associated with lower muscle strength, increased body instability, falls, and disability in older subjects [106, 107]. A significant association between the genotypes of the receptor for vitamin D with the strength of the quadriceps was also observed [108]. In addition, studies on vitamin D supplementation in elderly subjects with vitamin D deficiency showed an improvement in physical function and isometric knee extension versus placebo [109, 110]. In parallel with the decline in muscle mass and function with aging, there is a reduction of the expression of the receptors VDR (vitamin D receptor) in skeletal muscle [111]. Previous research has linked some VDR polymorphisms with the reduction of muscle mass and function in the elderly [112], suggesting that vitamin D plays a role in the development and progression of sarcopenia [113]. Prospective analysis of LASA (the Longitudinal Aging Study Amsterdam) has shown that low levels of 25OHD are predictive of an increased risk of recurrent falls at 1 year [114], reduction of muscle mass and strength in 3 years [115], and admission to nursing homes in six years [116]. Limited data suggest that physical activity or hypertrophy of skeletal muscle may be an important source of vitamin D [113]. It has been shown that skeletal muscle is the main deposit of vitamin D in infant rats [117]. The outdoor exercise improves the levels of 25OHD in the elderly [118]. This suggests that physical activity may influence muscle hypertrophy independently of 25OHD. Pfeifer et al. (2002) suggested that the muscle building exercises can increase the levels of 25OHD [119], but further studies are needed [113]. Studies conducted to evaluate the effects of vitamin D supplements on functional abilities are partially contradictory. Some studies have shown that vitamin D intake did not improve physical performance [120, 121]. These studies were conducted in subjects with normal vitamin D. Conversely, studies conducted in 122 elderly subjects with low levels of vitamin D have been shown to benefit significantly from supplementation of vitamin D. In particular, Dhesi et al. [122] demonstrated in a group of subjects (mean age 77 years), history of falls and blood values of vitamin D less than or equal to 12 micrograms/liter of a daily supplementation with 600 UI ergo-calciferol induced a 3% improvement of physical performance, as assessed by the "Aggregated Functional Performance Time (AFPT)" [123], while in the control group a 9% deterioration was reported. With respect to the postural stability, which is related to the state of vitamin D [124, 125], the study showed 13% improvement, while the control group had a worsening equal to 3%. In terms of reaction time, the treated group had a 13% improvement compared with a deterioration of 3% of the placebo group. However, it was not demonstrated an improvement in muscle strength [126].

Recently, Muir and Montero-Odasso [127] performed a meta-analysis on a collection of studies that those aged over 60 years participate in randomized control trials of the effect of vitamin D supplementation without an exercise intervention on muscle strength, gait, and balance. The meta-analysis suggests that vitamin D supplementation (800–1000 IU) daily was associated with improvements of muscle strength and balance.

In conclusion, in vitamin D deficient sarcopenic subjects, dietary vitamin D supplementation (800–1000 IU daily) could be promising and interesting for treatment of sarcopenia.

3.3.5 LONG-CHAIN OMEGA-3 FATTY ACIDS

This research has been carried out based on the keywords: "sarcopenia" AND "omega 3"; 10 articles were sourced. Among them, 2 RCT, 1 review, and 2 case control studies have been selected and discussed.

The ability of the skeletal musculature to use amino acids to build or renew constitutive proteins is gradually lost with age and this is partly due to a decline in skeletal muscle insulin sensitivity. Since long-chain omega-3 polyunsaturated fatty acids (LCn-3PUFA) from fish oil are known to improve insulin-mediated glucose metabolism in insulin-resistant states, some evidence in animal model suggests that polyunsaturated fatty acids might be a potentially useful therapeutic agent for the treatment and prevention of sarcopenia. It has been shown that providing feed enriched in fish oil to growing steers increases the activation (phosphorylation) of anabolic signaling proteins in muscle during administration of insulin and amino acids and increases the nonoxidative whole-body disposal of amino acids, an index of increased whole-body protein synthesis [128].

Furthermore, omega-3 fatty acid supplementation has been shown to prevent loss of muscle mass in burned guinea pigs [129].

In addition, omega-3 fatty acids present anti-inflammatory properties [130], which may also help alleviate the muscle anabolic resistance in older adults.

With respect to the effect of dietary omega-3 fatty acid supplementation on the rate of muscle protein synthesis and the anabolic signaling cascade in older adults, recently it has been demonstrated that omega-3 fatty acid supplementation had no effect on the basal rate of muscle protein synthesis but enhanced the hyperaminoacidemia-hyperinsulinemia-induced increase in the rate of muscle protein synthesis, which was accompanied by greater increases in muscle mTORSer2448 phosphorylation [131].

TABLE 4: Effect of nutrients or dietary supplementations on metabolic correlates of sarcopenia.

Nutrients or dietary supplementations	Recommendations	Specific effect
Proteins: average daily intake	It is recommended that the total protein intake should be 1–1.2 g/kg/day [16]	
Proteins: timing of intake	It is recommended to have 30 grams of protein of high biological value for each meal [25]	The elderly, compared with younger subjects, would require a larger amount of protein to obtain the same maximization of protein synthesis
Proteins: fast and slow	It is recommended to have whey protein ingestion because whey protein ingestion results in greater postprandial protein retention than does casein ingestion [31]	The greater anabolic properties of whey than of casein are mainly attributed to the faster digestion and absorption kinetics of whey, which results in a greater increase in postprandial plasma amino acid availability and thereby further stimulates muscle protein synthesis. Moreover, whey has a considerably higher leucine content
Proteins: animal and vegetal sources	When the total protein intake is adequate, the source of protein consumed (vegetal or animal) does not influence muscle strength and size [36]	Increases in muscle strength and size were not influenced by the predominant source of protein consumed by older men with adequate total protein intake
Branched chain amino acids (BCAAs),	It is recommended to have an adequate daily leucine supplementation (3 g/day)	A high proportion of leucine is required for optimal stimulation of the rate of muscle protein synthesis by essential amino acids in the elderly
Beta-hydroxy-methylbutyrate (HMB)	It is recommended to have a daily intake of beta-hydroxy butyrate (HMBb, 2 g/day) because it can attenuate the loss of muscle mass and increase muscle mass and strength [50]	Beta-hydroxy-beta-methylbutyrate is a product of leucine metabolism that has been shown to slow protein breakdown in muscle tissue
Creatine	It is recommended to have an adequate creatine supplementation because it could represent an intriguing intervention to counteract sarcopenia and in particular fatigue associated with sarcopenia; the timing of creatine ingestion (i.e., 0.03–0.5 g/kg before and after the sessions of resistance training) can be more relevant than the amount of creatine [73, 76]	The ingestion of an adequate creatine supplementation determines the increase in muscle phosphocreatine (PCr) and the energy provided for the phosphorylation of adenosine diphosphate (ADP) to adenosine triphosphate (ATP) during and after intense exercise largely depends on the amount of PCr stored in the muscle

TABLE 4: *Cont.*

Nutrients or dietary supplementations	Recommendations	Specific effect
Vitamin D	It is recommended to have a dietary vitamin D supplementation (800–1000 UI ergo-calciferol/day) in vitamin D deficient sarcopenic subjects [127]	Dietary vitamin D supplementation determines an increase of the expression of the receptors VDR (vitamin D receptor) in skeletal muscle
Antioxidants. vitamin E, vitamin C, carotenoids, and resveratrol	It is recommended to have a diet with high intake of fruits, vegetables whole grains, which is rich in antioxidant, and lower consumption of red meat and saturated fats, because it is associated with a reduced risk of inflammation correlated to oxidative damage [83]	Adherence to the diet rich in antioxidants is associated with lower circulating IL-6
Long-chain omega-3 polyunsaturated fatty acids (LC-3PUFA)	It is recommended to have dietary long-chain omega-3 polyunsaturated fatty acids (1.86 g eicosapentaenoic acid and 1.50 g docosahexaenoic acid/day) supplementation [131]	Long-chain omega-3 polyunsaturated fatty acids (LC-3PUFA) supplementation improves insulin-mediated glucose metabolism in insulin-resistant states and increases the activation (phosphorylation) of anabolic signaling proteins in muscle during administration of insulin and amino acids and increases the non-oxidative whole-body disposal of amino acids, an index of increased whole-body protein synthesis

In conclusion, dietary LCn-3PUFA supplementation could potentially provide a safe, simple, and low-cost intervention to counteract anabolic resistance and sarcopenia.

3.4 DISCUSSION

The more rational approach to delay the progression of sarcopenia is based on the combination of proper nutrition, possibly associated with the use of dietary supplements and a regular exercise program. An adequate intake of proteins (1.2/g/kg/day) is essential to prevent sarcopenia and aminoacids supplementation, in particular branched chain amino acids, is a well documented intervention for treating sarcopenia. Moreover, the current literature suggests that dietary LCn-3PUFA, vitamin D, and creatine supplementation could potentially provide a safe, simple, and low-cost intervention to counteract anabolic resistance and sarcopenia.

The author's nutritional recommendations have been showed in Table 4.

REFERENCES

1. A. J. Cruz-Jentoft, J. P. Baeyens, J. M. Bauer, et al., "Sarcopenia: European consensus on definition and diagnosis: report of the European working group on Sarcopenia in older people," Age and Ageing, vol. 39, no. 4, Article ID afq034, pp. 412–423, 2010.
2. G. A. van Kan, "Epidemiology and consequences of sarcopenia," The Journal of Nutrition, Health and Aging, vol. 13, no. 8, pp. 708–712, 2009.
3. http://www.iofbonehealth.org/.
4. W. R. Frontera, V. A. Hughes, R. A. Fielding, M. A. Fiatarone, W. J. Evans, and R. Roubenoff, "Aging of skeletal muscle: a 12-yr longitudinal study," The Journal of Applied Physiology, vol. 88, no. 4, pp. 1321–1326, 2000.
5. M. J. Delmonico, T. B. Harris, J. S. Lee et al., "Alternative definitions of sarcopenia, lower extremity performance, and functional impairment with aging in older men and women," Journal of the American Geriatrics Society, vol. 55, no. 5, pp. 769–774, 2007.
6. D. Dardevet, D. Rémond, M.-A. Peyron, I. Papet, I. Savary-Auzeloux, and L. Mosoni, "Muscle wasting and resistance of muscle anabolism: the "anabolic threshold concept" for adapted nutritional strategies during sarcopenia," The Scientific World Journal, vol. 2012, Article ID 269531, 6 pages, 2012.
7. V. Kumar, A. K. Abbas, and J. C. Aster, Eds., Robbins & Cotran Pathologic Basis of Disease, 2010.
8. U. Tarantino, G. Cannata, D. Lecce, M. Celi, I. Cerocchi, and R. Iundusi, "Incidence of fragility fractures," Aging Clinical and Experimental Research, vol. 19, no. 4, pp. 7–11, 2007.
9. V. Di Francesco, M. Zamboni, A. Dioli et al., "Delayed postprandial gastric emptying and impaired gallbladder contraction together with elevated cholecystokinin and peptide YY serum levels sustain satiety and inhibit hunger in healthy elderly persons," Journals of Gerontology—Series A Biological Sciences and Medical Sciences, vol. 60, no. 12, pp. 1581–1585, 2005.
10. V. di Francesco, M. Zamboni, E. Zoico et al., "Unbalanced serum leptin and ghrelin dynamics prolong postprandial satiety and inhibit hunger in healthy elderly: another

reason for the 'anorexia of aging'," The American Journal of Clinical Nutrition, vol. 83, no. 5, pp. 1149–1152, 2006.

11. H. Schröder, J. Marrugat, and M. I. Covas, "High monetary costs of dietary patterns associated with lower body mass index: a population-based study," International Journal of Obesity, vol. 30, no. 10, pp. 1574–1579, 2006.

12. P. Y. Cousson, M. Bessadet, E. Nicolas, J.-L. Veyrune, B. Lesourd, and C. Lassauzay, "Nutritional status, dietary intake and oral quality of life in elderly complete denture wearers," Gerodontology, vol. 29, no. 2, pp. e685–e692, 2012.

13. S. Iuliano, A. Olden, and J. Woods, "Meeting the nutritional needs of elderly residents in aged-care: are we doing enough?" The Journal of Nutrition Health and Aging, vol. 17, no. 6, pp. 503–508, 2013.

14. D. Volkert and C. C. Sieber, "Protein requirements in the elderly," International Journal for Vitamin and Nutrition Research, vol. 81, no. 2-3, pp. 109–119, 2011.

15. E. Smit, K. M. Winters-Stone, P. D. Loprinzi, A. M. Tang, and C. J. Crespo, "Lower nutritional status and higher food insufficiency in frail older US adults," British Journal of Nutrition, vol. 110, no. 1, pp. 172–178, 2013.

16. J. Bauer, G. Biolo, T. Cederholm et al., "Evidence-based recommendations for optimal dietary protein intake in older people: a position paper from the PROT-AGE Study Group," Journal of the American Medical Directors Association, vol. 14, no. 8, pp. 542–559, 2013.

17. J. E. Morley, J. M. Argiles, W. J. Evans et al., "Nutritional recommendations for the management of sarcopenia," Journal of the American Medical Directors Association, vol. 11, no. 6, pp. 391–396, 2010.

18. M. G. Giannoulis, F. C. Martin, K. S. Nair, A. M. Umpleby, and P. Sonksen, "Hormone replacement therapy and physical function in healthy older men. Time to talk hormones?" Endocrine Reviews, vol. 33, no. 3, pp. 314–377, 2012.

19. S. von Haehling, J. E. Morley, and S. D. Anker, "From muscle wasting to sarcopenia and myopenia: update 2012," Journal of Cachexia, Sarcopenia and Muscle, vol. 3, no. 4, pp. 213–217, 2012.

20. M. Egger, K. Dickersin, and G. D. Smith, "Problems and limitations in conducting systematic reviews," in Systematic Reviews in Health Care: Meta-Analysis in Context, M. Egger, G. D. Smith, and D. G. Altman, Eds., BMJ Books, London, UK, 2nd edition, 2001.

21. D. J. Millward, "Sufficient protein for our elders?" The American Journal of Clinical Nutrition, vol. 88, no. 5, pp. 1187–1188, 2008.

22. D. R. Moore, J. E. Tang, N. A. Burd, T. Rerecich, M. A. Tarnopolsky, and S. M. Phillips, "Differential stimulation of myofibrillar and sarcoplasmic protein synthesis with protein ingestion at rest and after resistance exercise," Journal of Physiology, vol. 587, no. 4, pp. 897–904, 2009.

23. J. Bohé, A. Low, R. R. Wolfe, and M. J. Rennie, "Human muscle protein synthesis is modulated by extracellular, not intramuscular amino acid availability: a dose-response study," Journal of Physiology, vol. 552, no. 1, pp. 315–324, 2003.

24. B. Pennings, B. Groen, A. de Lange, et al., "Amino acid absorption and subsequent muscle protein accretion following graded intakes of whey protein in elderly men," American Journal of Physiology: Endocrinology and Metabolism, vol. 302, no. 8, pp. E992–E999, 2012.

25. D. Paddon-Jones and B. B. Rasmussen, "Dietary protein recommendations and the prevention of sarcopenia," Current Opinion in Clinical Nutrition & Metabolic Care, vol. 12, no. 1, pp. 86–90, 2009.

26. Y. Boirie, M. Dangin, P. Gachon, M.-P. Vasson, J.-L. Maubois, and B. Beaufrère, "Slow and fast dietary proteins differently modulate postprandial protein accretion," Proceedings of the National Academy of Sciences of the United States of America, vol. 94, no. 26, pp. 14930–14935, 1997.

27. M. Dangin, Y. Boirie, C. Garcia-Rodenas et al., "The digestion rate of protein is an independent regulating factor of postprandial protein retention," American Journal of Physiology: Endocrinology and Metabolism, vol. 280, no. 2, pp. E340–E348, 2001.

28. R. Koopman, N. Crombach, A. P. Gijsen et al., "Ingestion of a protein hydrolysate is accompanied by an accelerated in vivo digestion and absorption rate when compared with its intact protein," The American Journal of Clinical Nutrition, vol. 90, no. 1, pp. 106–115, 2009.

29. M. Dangin, Y. Boirie, C. Guillet, and B. Beaufrère, "Influence of the protein digestion rate on protein turnover in young and elderly subjects," Journal of Nutrition, vol. 132, no. 10, pp. 3228S–3233S, 2002.

30. M. Dangin, C. Guillet, C. Garcia-Rodenas, et al., "The rate of protein digestion affects protein gain differently during aging in humans," The Journal of Physiology, vol. 549, no. 2, pp. 635–644, 2003.

31. B. Pennings, Y. Boirie, J. M. G. Senden, A. P. Gijsen, H. Kuipers, and L. J. C. Van Loon, "Whey protein stimulates postprandial muscle protein accretion more effectively than do casein and casein hydrolysate in older men," The American Journal of Clinical Nutrition, vol. 93, no. 5, pp. 997–1005, 2011.

32. J. E. Tang, D. R. Moore, G. W. Kujbida, M. A. Tarnopolsky, and S. M. Phillips, "Ingestion of whey hydrolysate, casein, or soy protein isolate: effects on mixed muscle protein synthesis at rest and following resistance exercise in young men," Journal of Applied Physiology, vol. 107, no. 3, pp. 987–992, 2009.

33. E. Volpi, H. Kobayashi, M. Sheffield-Moore, B. Mittendorfer, and R. R. Wolfe, "Essential amino acids are primarily responsible for the amino acid stimulation of muscle protein anabolism in healthy elderly adults," The American Journal of Clinical Nutrition, vol. 78, no. 2, pp. 250–258, 2003.

34. W. W. Campbell, M. L. Barton Jr., D. Cyr-Campbell, et al., "Effects of an omnivorous diet compared with a lactoovovegetarian diet on resistance-training-induced changes in body composition and skeletal muscle in older men," The American Journal of Clinical Nutrition, vol. 70, no. 6, pp. 1032–1039, 1999.

35. D. L. Pannemans, A. J. Wagenmakers, K. R. Westerterp, G. Schaafsma, and D. Halliday, "Effect of protein source and quantity on protein metabolism in elderly women," The American Journal of Clinical Nutrition, vol. 68, no. 6, pp. 1228–1235, 1998.

36. M. D. Haub, A. M. Wells, M. A. Tarnopolsky, and W. W. Campbell, "Effect of protein source on resistive-training-induced changes in body composition and muscle size in older men," American Journal of Clinical Nutrition, vol. 76, no. 3, pp. 511–517, 2002.

37. C. P. Sharp and D. R. Pearson, "Amino acid supplements and recovery from high-intensity resistance training," Journal of Strength & Conditioning Research, vol. 24, no. 4, pp. 1125–1130, 2010.

38. R. Riazi, L. J. Wykes, R. O. Ball, and P. B. Pencharz, "The total branched-chain amino acid requirement in young healthy adult men determined by indicator amino acid oxidation by use of L-[1-13C]phenylalanine," The Journal of Nutrition, vol. 133, no. 5, pp. 1383–1389, 2003.

39. S. R. Kimball and L. S. Jefferson, "Signaling pathways and molecular mechanisms through which branched-chain amino acids mediate translational control of protein synthesis," Journal of Nutrition, vol. 136, no. 1, pp. 227S–231S, 2006.

40. J. S. Greiwe, G. Kwon, M. L. McDaniel, and C. F. Semenkovich, "Leucine and insulin activate p70 S6 kinase through different pathways in human skeletal muscle," The American Journal of Physiology—Endocrinology and Metabolism, vol. 281, no. 3, pp. E466–E471, 2001.

41. S. R. Kimball and L. S. Jefferson, "Molecular mechanisms through which amino acids mediate signaling through the mammalian target of rapamycin," Current Opinion in Clinical Nutrition & Metabolic Care, vol. 7, no. 1, pp. 39–44, 2004.

42. L. Deldicque, D. Theisen, and M. Francaux, "Regulation of mTOR by amino acids and resistance exercise in skeletal muscle," European Journal of Applied Physiology, vol. 94, no. 1-2, pp. 1–10, 2005.

43. World Health Organization, "Energy and protein requirements—Technical Report Series," vol. 724, 1985.

44. L. E. Norton, Leucine is a critical factor determining protein quantity and quality of a complete meal to initiate muscle protein synthesis [Ph.D. thesis], University of Illinois, 2010.

45. E. L. Glynn, C. S. Fry, M. J. Drummond, et al., "Excess leucine intake enhances muscle anabolic signaling but not net protein anabolism in young men and women," Journal of Nutrition, vol. 140, no. 11, pp. 1970–1976, 2010.

46. J. C. Floyd Jr., S. S. Fajans, S. Pek, C. A. Thiffault, R. F. Knopf, and J. W. Conn, "Synergistic effect of essential amino acids and glucose upon insulin secretion in man.," Diabetes, vol. 19, no. 2, pp. 109–115, 1970.

47. D. K. Layman, "The role of leucine in weight loss diets and glucose homeostasis," Journal of Nutrition, vol. 133, no. 1, pp. 261S–267S, 2003.

48. S. L. Nissen and N. N. Abumrad, "Nutritional role of the leucine metabolite β-hydroxy β-methylbutyrate (HMB)," Journal of Nutritional Biochemistry, vol. 8, no. 6, pp. 300–311, 1997.

49. N. E. Zanchi, F. Gerlinger-Romero, L. Guimarães-Ferreira, et al., "HMB supplementation: clinical and athletic performance-related effects and mechanisms of action," Amino Acids, vol. 40, no. 4, pp. 1015–1025, 2011.

50. P. Flakoll, R. Sharp, S. Baier, D. Levenhagen, C. Carr, and S. Nissen, "Effect of β-hydroxy-β-methylbutyrate, arginine, and lysine supplementation on strength, functionality, body composition, and protein metabolism in elderly women," Nutrition, vol. 20, no. 5, pp. 445–451, 2004.

51. D. G. Candow, "Sarcopenia: current theories and the potential beneficial effect of creatine application strategies," Biogerontology, vol. 12, no. 4, pp. 273–281, 2011.

52. S. A. Smith, S. J. Montain, R. P. Matott, G. P. Zientara, F. A. Jolesz, and R. A. Fielding, "Creatine supplementation and age influence muscle metabolism during exercise," Journal of Applied Physiology, vol. 85, no. 4, pp. 1349–1356, 1998.

53. W. W. Campbell, L. J. O. Joseph, S. L. Davey, D. Cyr-Campbell, R. A. Anderson, and W. J. Evans, "Effects of resistance training and chromium picolinate on body composition and skeletal muscle in older men," Journal of Applied Physiology, vol. 86, no. 1, pp. 29–39, 1999.

54. K. E. Conley, S. A. Jubrias, and P. C. Esselman, "Oxidative capacity and ageing in human muscle," Journal of Physiology, vol. 526, no. 1, pp. 203–210, 2000.

55. E. S. Rawson, P. M. Clarkson, T. B. Price, and M. P. Miles, "Differential response of muscle phosphocreatine to creatine supplementation in young and old subjects," Acta Physiologica Scandinavica, vol. 174, no. 1, pp. 57–65, 2002.

56. P. A. Tesch, A. Thorsson, and N. Fujitsuka, "Creatine phosphate in fiber types of skeletal muscle before and after exhaustive exercise," Journal of Applied Physiology, vol. 66, no. 4, pp. 1756–1759, 1989.

57. J. Lexell and C. C. Taylor, "Variability in muscle fibre areas in whole human quadriceps muscle: Effects of increasing age," Journal of Anatomy, vol. 174, pp. 239–249, 1991.

58. I. R. Lanza, D. E. Befroy, and J. A. Kent-Braun, "Age-related changes in ATP-producing pathways in human skeletal muscle in vivo," Journal of Applied Physiology, vol. 99, no. 5, pp. 1736–1744, 2005.

59. P.-O. Larsson and K. Mosbach, "Affinity precipitation of enzymes," FEBS Letters, vol. 98, no. 2, pp. 333–338, 1979.

60. A. Brose, G. Parise, and M. A. Tarnopolsky, "Creatine supplementation enhances isometric strength and body composition improvements following strength exercise training in older adults," Journals of Gerontology—Series A Biological Sciences and Medical Sciences, vol. 58, no. 1, pp. 11–19, 2003.

61. K. Vandenberghe, M. Goris, P. van Hecke, M. van Leemputte, L. Vangerven, and P. Hespel, "Long-term creatine intake is beneficial to muscle performance during resistance training," Journal of Applied Physiology, vol. 83, no. 6, pp. 2055–2063, 1997.

62. J. S. Volek, N. D. Duncan, S. A. Mazzetti, et al., "Performance and muscle fiber adaptations to creatine supplementation and heavy resistance training," Medicine & Science in Sports & Exercise, vol. 31, no. 8, pp. 1147–1156, 1999.

63. B. O. Eijnde, M. Van Leemputte, M. Goris et al., "Effects of creatine supplementation and exercise training on fitness in men 55–75 yr old," Journal of Applied Physiology, vol. 95, no. 2, pp. 818–828, 2003.

64. M. Wyss, S. Felber, D. Skladal, A. Koller, C. Kremser, and W. Sperl, "The therapeutic potential of oral creatine supplementation in muscle disease," Medical Hypotheses, vol. 51, no. 4, pp. 333–336, 1998.

65. M. Tarnopolsky, G. Parise, M.-H. Fu et al., "Acute and moderate-term creatine monohydrate supplementation does not affect creatine transporter mRNA or protein content in either young of elderly humans," Molecular and Cellular Biochemistry, vol. 244, no. 1-2, pp. 159–166, 2003.

66. E. S. Rawson, M. L. Wehnert, and P. M. Clarkson, "Effects of 30 days of creatine ingestion in older men," European Journal of Applied Physiology and Occupational Physiology, vol. 80, no. 2, pp. 139–144, 1999.

67. E. S. Rawson and P. M. Clarkson, "Acute creatine supplementation in older men," International Journal of Sports Medicine, vol. 21, no. 1, pp. 71–75, 2000.

68. J. B. Wiroth, S. Bermon, S. Andreï, E. Dalloz, X. Hébuterne, and C. Dolisi, "Effects of oral creatine supplementation on maximal pedalling performance in older adults," European Journal of Applied Physiology, vol. 84, no. 6, pp. 533–539, 2001.

69. L. A. Gotshalk, J. S. Volek, R. S. Staron, C. R. Denegar, F. C. Hagerman, and W. J. Kraemer, "Creatine supplementation improves muscular performance in older men," Medicine and Science in Sports and Exercise, vol. 34, no. 3, pp. 537–543, 2002.

70. J. R. Stout, B. S. Graves, J. T. Cramer et al., "Effects of creatine supplementation on the onset of neuromuscular fatigue threshold and muscle strength in elderly men and women (64–86 years)," Journal of Nutrition, Health and Aging, vol. 11, no. 6, pp. 459–464, 2007.

71. L. A. Gotshalk, W. J. Kraemer, M. A. G. Mendonca et al., "Creatine supplementation improves muscular performance in older women," European Journal of Applied Physiology, vol. 102, no. 2, pp. 223–231, 2008.

72. S. Cañete, A. F. San Juan, M. Pérez et al., "Does creatine supplementation improve functional capacity in elderly women?" Journal of Strength and Conditioning Research, vol. 20, no. 1, pp. 22–28, 2006.

73. D. G. Candow and P. D. Chilibeck, "Timing of creatine or protein supplementation and resistance training in the elderly," Applied Physiology, Nutrition and Metabolism, vol. 33, no. 1, pp. 184–190, 2008.

74. M. Tamopolsky, A. Zimmer, J. Paikin et al., "Creatine monohydrate and conjugated linoleic acid improve strength and body composition following resistance exercise in older adults," PLoS ONE, vol. 2, no. 10, article e991, 2007.

75. P. D. Chilibeck, M. J. Chrusch, K. E. Chad, K. S. Davison, and D. G. Burke, "Creatine monohydrate and resistance training increase bone mineral content and density in older men," The Journal of Nutrition Health and Aging, vol. 9, no. 5, pp. 352–355, 2005.

76. V. J. Dalbo, M. D. Roberts, C. M. Lockwood, P. S. Tucker, R. B. Kreider, and C. M. Kerksick, "The effects of age on skeletal muscle and the phosphocreatine energy system: can creatine supplementation help older adults," Dynamic Medicine, vol. 8, no. 1, article 6, 2009.

77. R. S. Schwartz, C. Weindruch, and R. Weindruch, "Interventions based on the possibility that oxidative stress contributes to sarcopenia," The Journals of Gerontology, Series A: Biological Sciences and Medical Sciences, vol. 50, pp. 157–161, 1995.

78. D. McKenzie, E. Bua, S. McKiernan, Z. Cao, J. Wanagat, and J. M. Aiken, "Mitochondrial DNA deletion mutations: a causal role in sarcopenia," European Journal of Biochemistry, vol. 269, no. 8, pp. 2010–2015, 2002.

79. E. Carmeli, R. Coleman, and A. Z. Reznick, "The biochemistry of aging muscle," Experimental Gerontology, vol. 37, no. 4, pp. 477–489, 2002.

80. J. E. Morley and R. N. Baumgartner, "Cytokine-related aging process," Journals of Gerontology Series A: Biological Sciences and Medical Sciences, vol. 59, no. 9, pp. M924–M929, 2004.

81. K. S. Krabbe, M. Pedersen, and H. Bruunsgaard, "Inflammatory mediators in the elderly," Experimental Gerontology, vol. 39, no. 5, pp. 687–699, 2004.

82. L. A. Schaap, S. M. F. Pluijm, D. J. H. Deeg, and M. Visser, "Inflammatory markers and loss of muscle mass (sarcopenia) and strength," The American Journal of Medicine, vol. 119, no. 6, pp. 526.e9–526.e17, 2006.

83. R. D. Semba, F. Lauretani, and L. Ferrucci, "Carotenoids as protection against sarcopenia in older adults," Archives of Biochemistry and Biophysics, vol. 458, no. 2, pp. 141–145, 2007.

84. R. D. Semba, C. Blaum, J. M. Guralnik, D. T. Moncrief, M. O. Ricks, and L. P. Fried, "Carotenoid and vitamin E status are associated with indicators of sarcopenia among older women living in the community," Aging Clinical and Experimental Research, vol. 15, no. 6, pp. 482–487, 2003.

85. M. Cesari, M. Pahor, B. Bartali, et al., "Antioxidants and physical performance in elderly persons: the Invecchiare in Chianti (InCHIANTI) study," The American Journal of Clinical Nutrition, vol. 79, no. 2, pp. 289–294, 2004.

86. C. Chrysohoou, D. B. Panagiotakos, C. Pitsavos, U. N. Das, and C. Stefanadis, "Adherence to the Mediterranean diet attenuates inflammation and coagulation process in healthy adults: the ATTICA study," Journal of the American College of Cardiology, vol. 44, no. 1, pp. 152–158, 2004.

87. K. Esposito, R. Marfella, M. Ciotola et al., "Effect of a Mediterranean-style diet on endothelial dysfunction and markers of vascular inflammation in the metabolic syndrome: a randomized trial," The Journal of the American Medical Association, vol. 292, no. 12, pp. 1440–1446, 2004.

88. B. Marzani, M. Balage, A. Vénien et al., "Antioxidant supplementation restores defective leucine stimulation of protein synthesis in skeletal muscle from old rats," Journal of Nutrition, vol. 138, no. 11, pp. 2205–2211, 2008.

89. M. J. Ryan, H. J. Dudash, M. Docherty et al., "Vitamin E and C supplementation reduces oxidative stress, improves antioxidant enzymes and positive muscle work in chronically loaded muscles of aged rats," Experimental Gerontology, vol. 45, no. 11, pp. 882–895, 2010.

90. A. Ble, A. Cherubini, S. Volpato, et al., "Lower plasma vitamin E levels are associated with the frailty syndrome: the InCHIANTI study," Journals of Gerontology A: Biological Sciences and Medical Sciences, vol. 61, no. 3, pp. 278–283, 2006.

91. M. Meydani, W. J. Evans, G. Handelman et al., "Protective effect of vitamin E on exercise-induced oxidative damage in young and older adults," The American Journal of Physiology—Regulatory Integrative and Comparative Physiology, vol. 264, no. 5, pp. R992–R998, 1993.

92. J. M. Sacheck, P. E. Milbury, J. G. Cannon, R. Roubenoff, and J. B. Blumberg, "Effect of vitamin E and eccentric exercise on selected biomarkers of oxidative stress in young and elderly men," Free Radical Biology and Medicine, vol. 34, no. 12, pp. 1575–1588, 2003.

93. S. A. Devi, S. Prathima, and M. V. V. Subramanyam, "Dietary vitamin E and physical exercise: I. Altered endurance capacity and plasma lipid profile in ageing rats," Experimental Gerontology, vol. 38, no. 3, pp. 285–290, 2003.

94. S. P. Lee, G. Y. Mar, and L. T. Ng, "Effects of tocotrienol-rich fraction on exercise endurance capacity and oxidative stress in forced swimming rats," European Journal of Applied Physiology, vol. 107, no. 5, pp. 587–595, 2009.

95. P. Lips, "Vitamin D deficiency and secondary hyperparathyroidism in the elderly: consequences for bone loss and fractures and therapeutic implications," Endocrine Reviews, vol. 22, no. 4, pp. 477–501, 2001.

96. J.-C. Souberbielle, C. Cormier, C. Kindermans et al., "Vitamin D status and redefining serum parathyroid hormone reference range in the elderly," Journal of Clinical Endocrinology and Metabolism, vol. 86, no. 7, pp. 3086–3090, 2001.

97. R. K. Skinner, S. Sherlock, R. G. Long, and M. R. Wills, "25-hydroxylation of vitamin D in primary biliary cirrhosis," The Lancet, vol. 1, no. 8014, pp. 720–721, 1977.

98. K. Farrington, R. K. Skinner, Z. Varghese, and J. F. Moorhead, "Hepatic metabolism of vitamin D in chronic renal failure," The Lancet, vol. 1, no. 8111, p. 321, 1979.

99. J. C. Gallagher, B. L. Riggs, J. Eisman, A. Hamstra, S. B. Arnaud, and H. F. DeLuca, "Intestinal calcium absorption and serum vitamin D metabolites in normal subjects and osteoporotic patients. Effect of age and dietary calcium," Journal of Clinical Investigation, vol. 64, no. 3, pp. 729–736, 1979.

100. H. Reichel and A. W. Norman, "Systemic effects of vitamin D," Annual Review of Medicine, vol. 40, pp. 71–78, 1989.

101. M. R. Walters, "Newly identified actions of the vitamin D endocrine system," Endocrine Research, vol. 18, no. 4, pp. 719–764, 1992.

102. H. A. Pols, J. C. Birkenhager, J. A. Foekens, and J. P. van Leeuwen, "Vitamin D: a modulator of cell proliferation and differentiation," The Journal of Steroid Biochemistry and Molecular Biology, vol. 37, no. 6, pp. 873–876, 1990.

103. C. Garland, R. B. Shekelle, E. Barrett-Connor, M. H. Criqui, A. H. Rossof, and O. Paul, "Dietary vitamin D and calcium and risk of colorectal cancer: a 19-year prospective study in men," The Lancet, vol. 1, no. 8424, pp. 307–309, 1985.

104. F. C. Garland, C. F. Garland, E. D. Gorham, and J. F. Young, "Geographic variation in breast cancer mortality in the United States: a hypothesis involving exposure to solar radiation," Preventive Medicine, vol. 19, no. 6, pp. 614–622, 1990.

105. B. L. Nyomba, R. Bouillon, and P. de Moor, "Influence of vitamin D status on insulin secretion and glucose tolerance in the rabbit," Endocrinology, vol. 115, no. 1, pp. 191–197, 1984.

106. H. A. Bischoff, H. B. Stahelin, N. Urscheler et al., "Muscle strength in the elderly: its relation to vitamin D metabolites," Archives of Physical Medicine and Rehabilitation, vol. 80, no. 1, pp. 54–58, 1999.

107. J. K. Dhesi, L. M. Bearne, C. Moniz et al., "Neuromuscular and psychomotor function in elderly subjects who fall and the relationship with vitamin D status," Journal of Bone and Mineral Research, vol. 17, no. 5, pp. 891–897, 2002.

108. P. Geusens, C. Vandevyver, J. Vanhoof, J.-J. Cassiman, S. Boonen, and J. Raus, "Quadriceps and grip strength are related to vitamin D receptor genotype in elderly nonobese women," Journal of Bone and Mineral Research, vol. 12, no. 12, pp. 2082–2088, 1997.

109. H. J. J. Verhaar, M. M. Samson, P. A. F. Jansen, P. L. de Vreede, J. W. Manten, and S. A. Duursma, "Muscle strength, functional mobility and vitamin D in older women," Aging Clinical and Experimental Research, vol. 12, no. 6, pp. 455–460, 2000.

110. F. M. Gloth III, C. M. Gundberg, B. W. Hollis, J. G. Haddad Jr., and J. D. Tobin, "Vitamin D deficiency in homebound elderly persons," Journal of the American Medical Association, vol. 274, no. 21, pp. 1683–1686, 1995.

111. H. A. Bischoff-Ferrari, M. Borchers, F. Gudat, U. Dürmüller, H. B. Stähelin, and W. Dick, "Vitamin D receptor expression in human muscle tissue decreases with age," Journal of Bone and Mineral Research, vol. 19, no. 2, pp. 265–269, 2004.

112. S. M. Roth, J. M. Zmuda, J. A. Cauley, P. R. Shea, and R. E. Ferrell, "Vitamin D receptor genotype is associated with fat-free mass and sarcopenia in elderly men," Journals of Gerontology Series A Biological Sciences and Medical Sciences, vol. 59, no. 1, pp. 10–15, 2004.

113. D. Scott, L. Blizzard, J. Fell, C. Ding, T. Winzenberg, and G. Jones, "A prospective study of the associations between 25-hydroxy-vitamin D, sarcopenia progression and physical activity in older adults," Clinical Endocrinology, vol. 73, no. 5, pp. 581–587, 2010.

114. M. B. Snijder, N. M. Van Schoor, S. M. F. Pluijm, R. M. Van Dam, M. Visser, and P. Lips, "Vitamin D status in relation to one-year risk of recurrent falling in older men and women," Journal of Clinical Endocrinology and Metabolism, vol. 91, no. 8, pp. 2980–2985, 2006.

115. M. Visser, D. J. H. Deeg, and P. Lips, "Low Vitamin D and high parathyroid hormone levels as determinants of loss of muscle strength and muscle mass (sarcopenia): the longitudinal aging study Amsterdam," Journal of Clinical Endocrinology and Metabolism, vol. 88, no. 12, pp. 5766–5772, 2003.

116. M. Visser, D. J. H. Deeg, M. T. E. Puts, J. C. Seidell, and P. Lips, "Low serum concentrations of 25-hydroxyvitamin D in older persons and the risk of nursing home admission," The American Journal of Clinical Nutrition, vol. 84, no. 3, pp. 616–622, 2006.

117. M. R. Clements and D. R. Fraser, "Vitamin D supply to the rat fetus and neonate," The Journal of Clinical Investigation, vol. 81, no. 6, pp. 1768–1773, 1988.

118. R. Scragg and C. A. Camargo Jr., "Frequency of leisure-time physical activity and serum 25-hydroxyvitamin D levels in the US population: results from the third national health and nutrition examination survey," American Journal of Epidemiology, vol. 168, no. 6, pp. 577–586, 2008.

119. M. Pfeifer, B. Begerow, and H. W. Minne, "Vitamin D and muscle function," Osteoporosis International, vol. 13, no. 3, pp. 187–194, 2002.

120. O. H. Sørensen, B. Lund, B. Saltin et al., "Myopathy in bone loss of ageing: improvement by treatment with 1α-hydroxycholecalciferol and calcium," Clinical Science, vol. 56, no. 2, pp. 157–161, 1979.

121. A. M. Kenny, B. Biskup, B. Robbins, G. Marcella, and J. A. Burleson, "Effects of vitamin D supplementation on strength, physical function, and health perception in older, community-dwelling men," Journal of the American Geriatrics Society, vol. 51, no. 12, pp. 1762–1767, 2003.

122. J. K. Dhesi, S. H. D. Jackson, L. M. Bearne et al., "Vitamin D supplementation improves neuromuscular function in older people who fall," Age and Ageing, vol. 33, no. 6, pp. 589–595, 2004.

123. M. V. Hurler, J. Rees, and D. J. Newham, "Quadriceps function, proprioceptive acuity and functional performance in healthy young, middle-aged and elderly subjects," Age and Ageing, vol. 27, no. 1, pp. 55–62, 1998.

124. S. R. Lord, P. N. Sambrook, C. Gilbert et al., "Postural stability, falls and fractures in the elderly: results from the dubbo osteoporosis epidemiology study," Medical Journal of Australia, vol. 160, no. 11, pp. 688–691, 1994.

125. T. Nguyen, P. Sambrook, P. Kelly et al., "Prediction of osteoporotic fractures by postural instability and bone density," British Medical Journal, vol. 307, no. 6912, pp. 1111–1115, 1993.

126. E. Sohl, R. T. de Jongh, A. C. Heijboer et al., "Vitamin D status is associated with physical performance: the results of three independent cohorts," Osteoporosis International, vol. 24, no. 1, pp. 187–196, 2013.

127. S. W. Muir and M. Montero-Odasso, "Effect of vitamin D supplementation on muscle strength, gait and balance in older adults: a systematic review and meta-analysis," Journal of the American Geriatrics Society, vol. 59, no. 12, pp. 2291–2300, 2011.

128. A.-A. Gingras, P. J. White, P. Y. Chouinard et al., "Long-chain omega-3 fatty acids regulate bovine whole-body protein metabolism by promoting muscle insulin signalling to the Akt-mTOR-S6K1 pathway and insulin sensitivity," Journal of Physiology, vol. 579, no. 1, pp. 269–284, 2007.

129. J. W. Alexander, H. Saito, O. Trocki, and C. K. Ogle, "The importance of lipid type in the diet after burn injury," Annals of Surgery, vol. 204, no. 1, pp. 1–8, 1986.

130. J. W. Fetterman Jr. and M. M. Zdanowicz, "Therapeutic potential of n-3 polyunsaturated fatty acids in disease," The American Journal of Health-System Pharmacy, vol. 66, no. 13, pp. 1169–1179, 2009.

131. G. I. Smith, P. Atherton, D. N. Reeds et al., "Dietary omega-3 fatty acid supplementation increases the rate of muscle protein synthesis in older adults: a randomized controlled trial," The American Journal of Clinical Nutrition, vol. 93, no. 2, pp. 402–412, 2011.

PART II

PROTEIN AND EXERCISE

CHAPTER 4

Whey Protein and Essential Amino Acids Promote the Reduction of Adipose Tissue and Increased Muscle Protein Synthesis During Caloric Restriction-Induced Weight Loss in Elderly, Obese Individuals

ROBERT H. COKER, SHARON MILLER, SCOTT SCHUTZLER, NICOLAAS DEUTZ, AND ROBERT R. WOLFE

4.1 INTRODUCTION

The elderly population is particularly prone to the accelerated loss of muscle mass or sarcopenia with advancing age. Sarcopenia is complicated by the combined influence of physical inactivity and poor nutrition that often accompanies the aging process [1]. The condition of sarcopenia is inextricably linked to loss of functional capacity, and increased risk of morbidity

Whey Protein and Essential Amino Acids Promote the Reduction of Adipose Tissue and Increased Muscle Protein Synthesis During Caloric Restriction-Induced Weight Loss in Elderly, Obese Individuals. © Coker RH, Miller S, Schutzler S, Deutz N, and Wolfe RR. Nutrition Journal 11,105 (2012), doi:10.1186/1475-2891-11-105. Licensed under a Creative Commons Attribution 2.0 Generic License, http://creativecommons.org/licenses/by/2.0/.

and mortality [2-5]. While the precise etiology of sarcopenia has yet to elucidated, recent evidence points to a overall reduction in the anabolic response of skeletal muscle to the consumption of mixed macronutrient intake [6,7], potentially contributing to chronic, progressive reduction of muscle mass with advancing age.

Combined with the challenge of sarcopenia, the prevalence of obesity among elderly adults has increased dramatically [8]. The concomitant development of sarcopenic obesity results in the formation of overlapping health problems that combine the negative influences of inflammation, frailty and metabolic syndrome [9,10]. Moreover, caloric restriction-induced weight loss that is often utilized to dampen the negative consequences of metabolic syndrome may be considered counterproductive when measured against weight loss-induced acceleration of sarcopenia [11].

While moderate increases in dietary protein intake beyond the Recommended Daily Allowance (RDA) have been suggested to enhance anabolic activity in skeletal muscle [12], amino acid consumption have also been shown to facilitate an increase in muscle protein synthesis [13,14]. In addition, supplementation of high quality protein twice a day promoted an increase in lean body mass, strength and functional capacity in elderly volunteers with concomitant changes in diet or activity [15]. The development of the most potent formulation of whey protein and amino acid supplementation would potentially protect against the negative consequences of sarcopenic obesity in the elderly.

To meet our aims, we chose a caloric restriction paradigm designed to elicit 7% weight loss and decrease metabolic risk [15]. Our overriding hypothesis was that the supplementation of our whey protein + essential amino acid meal replacement (EAAMR) formulation will be required to promote the optimal reduction of adipose tissue while preserving lean tissue in obese, elderly adults. We also proposed that the changes in body composition would occur in conjunction with an increase in the skeletal muscle protein fractional synthetic rate (FSR) to EAAMR compared to competitive meal replacement (CMR).

4.1 METHODS

4.1.1 RECRUITMENT AND SCREENING

We recruited and enrolled healthy males and females, ages = 65-80 years of age and of all races and ethnic backgrounds for a caloric restriction-based weight loss intervention designed to promote 7% weight loss. All volunteers were weight stable and were not participating in a weight loss or exercise program at the time of enrollment. Volunteers refrained from alcohol consumption (24 h) and intense physical activities (72 h) prior to pre- and post-weight loss study visits. Volunteers also reported in the fasted state beginning at 20:00 h the day prior to pre- and post-weight loss study visits.

In conjunction with the consumption of solid food in both groups, subjects were randomized to one of two groups that would consume either EAAMR or a CMR five times/day. Volunteers were recruited through the conspicuous placement of IRB-approved flyer throughout the community, and completed a telephone screening. Potential subjects were then asked to report for a visit to our clinic, where they were then consented.

We excluded any patient with a chronic inflammatory condition or malignancy. We also excluded any individual who could not discontinue the use of aspirin or other anti-coagulant medication. In addition, volunteers with a history of diabetes, cardiovascular disease or recent history of cancer (within 3 years) were also excluded. We also excluded any individual whose body weight had changed more than 10 pounds in the previous 3 months. Due to the nature of the weight loss intervention, we excluded all volunteer with allergies to milk products.

It would have been impractical to exclude all patients taking medications, and we could not ethically discontinue all medications for the duration of the intervention. As a result, we permitted the concomitant use of HMG CoA-reductase inhibitors, thiazide diuretics, beta-1 selective blockers, clonidine, antidepressant selective serotonin reuptake inhibitors, and estrogen replacement/oral contraceptives.

4.1.2 WEIGHT LOSS STUDY DESIGN

All medical screenings included a baseline blood sample collection for a lipid and metabolic panel. Based on eligibility, volunteers were randomly assigned to one of two groups: 1) 7% weight loss with EAAMR, or 7% weight loss using CMR. Both groups consumed 5 servings of a meal replacement (170 kcal x 5 servings/day) and ~400 kcal/day of solid food that yielded ~1250 kcal/day. Regardless of group assignment, the total kcal/week remained constant throughout the intervention to ensure a high level of compliance and allow for consistent, reliable reductions in the amount of weight lost. Obese, elderly volunteers in both groups (EAAMR or CMR) completed one visits that included a stable isotope experimental session described below. Following the completion of the experimental testing sessions, all volunteers were instructed to consume a low calorie diet for 8 weeks. During this weight loss intervention, volunteers consumed ~1250 kcal/day consisting of solid food and liquid supplements. Although the time course of the caloric restriction phase may seem somewhat overbearing from a logistical standpoint, the difficulty entailed was dramatically reduced through the use of pre-packaged meal replacements either in the form of EAAMR or CMR that required no preparation. In addition, we suggested solid food products preferred by clients in the UAMS weight loss clinic that achieve optimal compliance and consistently provide predictable weight loss. Body weight and body composition was measured weekly by our clinical staff. Any volunteer that failed to lose weight on two subsequent visits was removed from the study for noncompliance to the protocol.

Experimental Paradigm. Prior to initiation of the weight loss intervention, volunteers in both groups completed a testing session designed to examine the influence of EAAMR and CMR on skeletal muscle protein FSR. Subjects were instructed to fast after 2200 hrs, with water permitted. On the day of the testing session, subjects reported to the study site, and two intravenous catheters were inserted by our registered nurse. One catheter was utilized for the infusion of stable isotopes, while the other was utilized for blood sampling. A primed (2 μmol/kg), constant (0.07 μmol/kg/min) infusion of ring- [13]C6-phenylalanine began after obtaining a background blood sample and the isotopic infusion was maintained for 6 hours. Blood was sampled at 20 min intervals to determine tracer enrichment and blood amino acid levels. A tissue biopsy of the vastus lateralis muscle was taken under local anesthesia at t=60, 180 and 240 min [16]. Biopsies were utilized for the determination of skeletal muscle protein FSR by tracer methodology [17]. After the initial biopsy at t=60 min, subjects were provided with a single dose of EAAMR or CMR. At t=180 and t=240 min, two biopsies were taken to evaluate the influence of EAAMR or CMR on the incorporation of the phenyalalnine tracer into the muscle.

We have previously demonstrated that characterization of this acute response to EAA reflects the 24 h response, and that skeletal muscle protein FSR remains constant for the entire 8 hours if no drink is given [18]. While both products were presented in a tetrapak, and in a palatability format that precluded specific determination of either product by the volunteer, the knowledge of the product alone would not have influenced skeletal muscle protein FSR. Following the completion of the infusion protocol and biopsy, catheters were withdrawn, the biopsy site was dressed, and after consult with the study physician, the subject was released.

4.1.3 BODY WEIGHT AND COMPOSITION

We measured body mass, BMI, and body composition pre- and post-weight loss, and during each weekly visit. Total body mass was to the nearest 0.1 kg using an electronic scale (Ohaus Corp, USA). A GAIA 359 PLUS (Kyungsang Buk-do, KOREA) unit measured fat mass, lean tissue mass, soft lean mass and percent body fat using bioelectrical impedance analysis via the tetra-polar electrode method.

4.1.4 ANALYSIS OF SAMPLES

4.1.4.1 BLOOD

The isotopic enrichment of L-[ring- [13]C_6phenylalanine in blood was determined on an HP Model 5973 GCMS (Hewlett-Packard Co., Palo Alto, CA) by electron impact ionization and selected ion monitoring [17].

4.1.4.2 MUSCLE

Tissue biopsies of the vastus lateralis were immediately rinsed with cold saline, blotted, and frozen in liquid nitrogen. These tissue samples were stored at -80°C until processed. The TBDMS derivative was prepared for the intracellular free water as previously described 56, and analyzed by gas chromatography/mass spectrometry (GCMS) (Model 5973) using electron impact ionization. The protein-bound enrichment of phenylalanine was analyzed as previously described by GCMS [17].

4.1.4.3 CALCULATION OF SKELETAL MUSCLE PROTEIN FSR

Skeletal muscle protein FSR was calculated from the determination of the rate of tracer incorporation into the protein and the enrichment of the intracellular pool as the precursor:

$$FSR = [(E_{p2} - E_{p1})/(E_{M}t)] \cdot 60 \cdot 100$$

where E_{p1} and E_{p2} are the enrichments of the protein-bound L-[ring- [13] C_{6}phenylalanine at t=60, 180, and 240 min. E_{M} represented the average intracellular L-[ring- [13]C_{6}phenylalanine enrichments over the time of incorporation; and t is the time in minutes. The factors 60 and 100 were required to express skeletal muscle protein FSR in percent per hour. Skeletal muscle protein FSR was calculated from the biopsies at 60 and 180 min to determine fasting protein synthetic rate, and also from t=180 min and t=240 min to determine the effect of drink administration. The drink was given immediately after the 180 min biopsy and thus incorporation of the tracer from 180-240 min reflects acute nutritional response. From the skeletal muscle protein FSR and lean body mass determination the total change in skeletal muscle protein synthesis was estimated [18].

TABLE 1: Age and Anthropometrics (pre- and post- intervention

	EAAMR		CMR	
	Pre WL	Post WL	Pre WL	Post WL
Age	70±2		68±2	
Body weight (kg)	92.3±3.9	84.5±3.3*	91.4±2.4	84.6±2.4
Body mass index (kg/m²)	31.3±0.5	29.1±0.7	31.3±0.5	30.0±0.4
Percent Fat	41.8±1.1	36.3±1.1*	38.9±1.5	37.5±0.2

Values are means±SEM. There were no significant differences in these variables at basekine. *Denotes a significant change from pre-to post-intervention (P<0.05).

4.1.5 STATISTICAL ANALYSIS

Variables were summarized using means, standard deviations, medians and ranges. The groups were compared with respect to changes in adipose and lean tissue, and skeletal muscle protein synthesis as measured by skeletal muscle protein FSR, using Student two-sample t-tests. An alpha-level of 5% was used to determine statistical significance.

4.2 RESULTS

4.2.1 SUBJECTS

We enrolled 12 volunteers, and one subject dropped out due to their inability to comply with the protocol. Therefore, 11 obese women and men lost 7% of total body weight, and completed all aspects of the study. There were no baseline differences in age, BMI or percent fat, or the distribution of gender equity between the EAAMR and CMR groups (p>0.05) (Table 1). Of these individuals, one of the individuals in the EAAMR group had undergone a hip replacement in 2003, and one of the individuals in the CMR group had been successfully treated for cancer in 1999. In addition, four out of the 11 individuals were taking simvastatin or nexium, and two of the 11 individuals were taking estradiol.

Baseline Metabolic and Lipid Parameters. There were no significant differences in the fasting glucose values between EAAMR and CMR (p>0.05) (Table 2). In addition, there were no differences in triglycerides, total cholesterol, high density lipoprotein or very low density lipoprotein (p>0.05). While there was a trend towards reduced low density lipoprotein in EAAMR, it was not significant (p=0.10).

TABLE 2: Plasma glucose and lipid variables

	EAAMR	CMR
Glucose (mg/dl)	97±4	103±7
Triglycerides (mg/dl)	154±22	115±13
Total cholesterol (mg/dl)	168±7	191±9
High density lipoprotein (mg/dl)	47±4	47±5
Very low density lipoprotein	31±4	23±3
Low density lipoprotein (mg/dl)	86±9	121±12

Values are means±SEM. There were no significant differences in these baseline variables (P>0.05).

4.2.2 ANTHROPOMETRICS

Relative to the caloric restriction-induced weight loss interventions, there was a significant and similar 7% reduction in body weight in EAAMR and CMR (Table 1). While there was no difference in absolute amount of weight loss, EAAMR promoted a greater reduction in adipose tissue compared to CMR (p<0.05) (Table 1) (Figure 1). The consumption of the EAAMR also seemed to foster greater preservation of lean tissue compared to CMR. However, the difference between EAAMR and CMR was not significant (p=0.26) (Figure 2).

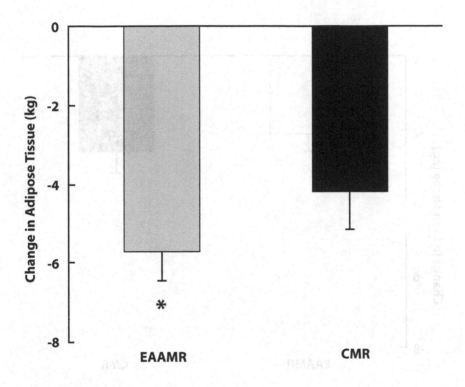

FIGURE 1: Change in adipose tissue in EAAMR and CMR following weight loss. * Denotes significance difference between groups.

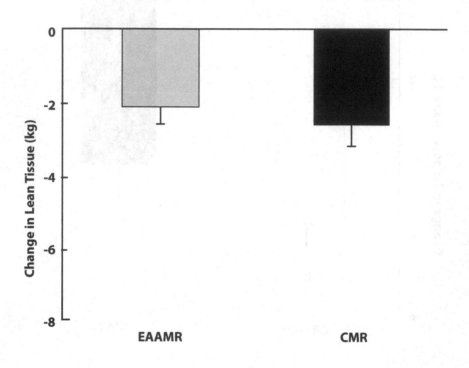

FIGURE 2: Change in lean tissue in EAAMR and CMR following weight loss.

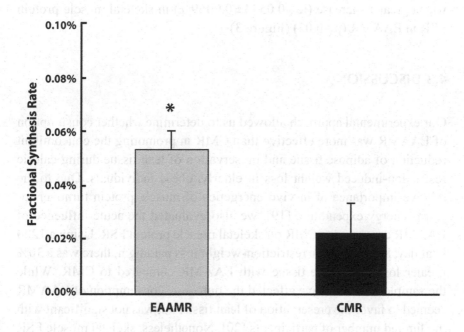

FIGURE 3: Change in feeding induced skeletal muscle FSR in EAAMR and CMR following weight loss. * Denotes significance difference between groups.

4.2.3 PLASMA AND MUSCLE PHENYALANINE ENRICHMENTS

Basal plasma phenyalanine enrichments were not different between EAAMR and CMR ($p > 0.05$). Upon ingestion of each product, muscle phenyalanine enrichments increased ($p < 0.05$) dramatically in EAAMR compared to no change with CMR ($p > 0.05$).

Skeletal Muscle Fractional Protein Synthesis. Skeletal muscle protein FSR increased by $0.0142 \pm 0.0154\%$ in CMR (Figure 3). In contrast, there was a greater increase (ie., $0.0534 \pm 0.0069\%$) in skeletal muscle protein FSR in EAAMR ($p < 0.05$) (Figure 3).

4.3 DISCUSSION

Our experimental approach allowed us to determine whether consumption of EAAMR was more effective than CMR in promoting the concomitant reduction of adipose tissue and preservation of lean tissue during caloric restriction-induced weight loss in elderly, obese individuals. Due to the relative importance of in vivo energetics of muscle protein turnover towards energy expenditure [19], we also evaluated the acute influence of EAAMR compared to CMR on skeletal muscle protein FSR. Using a 1200 kcal/day, 8 week caloric restriction-weight loss paradigm, there was a 30% greater loss of adipose tissue with EAAMR compared to CMR. While the sample size may have affected the outcome, consumption of EAAMR seemed to favor the preservation of lean tissue but was not significant with the limited number of participants [20]. Nonetheless, skeletal muscle FSR was significantly higher with the acute consumption of EAAMR versus CMR, and may have been responsible for the preferential loss of adipose tissue with EAAMR.

The macronutrients contained in a single serving of EAAMR and CMR are described in Table 3. It is important to mention that the total caloric value for a single serving of EAAMR and CMR was ~170 kcal. The amount/type of carbohydrate was identical in both formulations. Also, the total amount of fat was 4 grams in EAAMR and 3 grams in CMR with the proportion of saturated fat being 1 gram in both EAAMR and CMR. The primary differences between EAAMR and CMR were the total amount/

type of intact protein and the total amount and formulation of essential amino acids. While EAAMR contained 7 grams of intact protein from whey protein, CMR contained 14 grams intact protein derived from sodium casenate and calcium casenate. Leucine comprises approximately 40% of the EAAs, in accord with our research defining the optimal formulation of EAAs for elderly [13,16,21]. We have also previously demonstrated greater accrual of muscle protein with whey protein [22], and we proposed that the combination of whey protein plus 6 grams of our proprietary essential amino acid formulation would be the most effective approach against sarcopenic obesity. The total amount of essential amino acids derived from intact protein and the essential amino formulation was 9.7 grams in EAAMR. The total amount of essential amino acids derived from intact protein was 3.5 grams in CMR. The description highlights the fundamental differences between EAAMR and CMR and the rationale for the source of intact protein and importance of the essential amino acids formulation.

TABLE 3: Nutrition Facts for EAAMR and CMR

	EAAMR	CMR
Calories	170	170
Total Fat (grams)	4	3
Saturated Fat (grams)	1	1
Trans Fat (grams)	0	0
Cholesterol (mg)	5	5
Sodium (mg)	220	220
Potassium (mg)	460	460
Total Carbohydrate	22	22
Fiber (grams)	1	1
Sugars (grams)	17	17
Protein (grams)	7	14
Essential amino acid formulation	6	0

There are several factors that should be considered when comparing the effects of meal replacements on changes in body composition and the energetics of skeletal muscle in the context of weight loss. Typically, total energy expenditure is represented by the sum of resting energy expenditure, the thermic effect of food, and the energy related to activity [23]. Given that that the experimental protocol was identical in both groups except for the nutrient profile of the meal replacements, we can make the assumption that basal energy expenditure and activity related energy expenditure were not responsible for the differences between EAAMR and CMR. Other factors such as the administration of the meal replacements, timing sequence of the protocol, the characterization of the subjects, and the length of the protocol could have also influenced the results related to body composition. These factors were either well controlled by the experimental design or were not significantly different between groups. Therefore, the impact of the nutrient profiles (ie., type of protein and formulation of essential amino acids) was likely responsible for the greater loss of adipose tissue through differences in diet-induced energy expenditure or the impact of muscle loss sparing on overall energy expenditure.

While we were not able to achieve statistical significance in terms of EAAMR promoting enhanced preservation of lean tissue, the sparing influence of muscle loss might have been demonstrated with a larger sample size. This remains an important issue. Acute administration of EAAMR did promote a significant increase in skeletal muscle protein FSR compared to CMR. Assuming that the energy cost of protein synthesis is 3.6 kJ/g and the baseline GAIA-derived lean tissue mass was 56.4 kg for EAAMR and 54.4 kg for the CMR, we can extrapolate that the overall energy discrepancy between the two groups was roughly equivalent to 27,170 kcal or 3.5 kg of weight loss across the entire caloric restriction-based weight loss paradigm. Based on the amount of total lean mass in each group, this value takes into account a consistent intervention structure of five servings/day across an eight week period. In short, these calculations suggest that differences in the source of intact protein/formulation of EAA may have a significant influence on diet induced energy expenditure that coincides closely with the greater reduction of adipose tissue in EAAMR compared to CMR.

Despite the greater stimulation of skeletal muscle protein FSR in EAAMR, these individuals still lost 2.1 ± 0.5 kg of lean tissue. With an estimated protein intake of 90 grams/day, this might seem confusing since the recommended dietary allowance (RDA) of 0.88 kg^{-1}· d^{-1} would equate to 50 grams of protein in the EAAMR group [24]. On the other hand, the RDA for protein has been shown to be inadequate for the maintenance of lean body mass in the elderly population [25]. Therefore, the optimal amount of protein intake in the elderly population remains uncertain [26], especially in the context of caloric restriction. The underlying reason for the lack of precise information when it comes to protein intake during weight loss in this population may be largely due to the fact that the protein RDA levels were determined by studies using nitrogen balance in younger individuals [27]. In order to attenuate the acceleration of sarcopenia, the amount/type of protein and/or formulation of essential amino acids must be elucidated to promote efficacious caloric restriction-based weight loss in the elderly.

For the effective treatment of sarcopenic obesity, it might be argued that exercise training should be utilized instead of or in conjunction with caloric restriction [28]. In fact, we have also recently demonstrated greater reduction of adipose tissue through the use of exercise-induced weight loss compared to caloric restriction [29]. There was also a greater preservation of lean tissue with exercise-induced weight loss compared to caloric restriction-induced weight loss [30]. Unfortunately, individuals at risk for metabolic disease may be limited in their ability to even meet the guidelines for physical activity much less induce weight loss through exercise training [30]. In fact, only 30% of young, healthy Americans currently utilize exercise as an efficacious strategy for weight loss [31]. These issues are compounded by the difficulties entailed with maintaining exercise compliance [32]. Therefore, caloric restriction-based weight loss provides a potential avenue for an improvement in metabolic health status and may influence functional status as well due to the inverse relationship between obesity and physical activity [33].

Weight loss in the obese, elderly population is typically contraindicated due to difficult, clinical conundrum of sarcopenic obesity. While the results of the present study do not suggest that EAAMR would be sufficient to completely prevent the reduction of lean tissue during caloric restriction-induced weight loss, we were able to demonstrate the preferential reduction of adipose tissue with 7% weight loss. Moreover, the acute anabolic response to the ingestion of EAAMR was directly linked to increased diet-induced caloric expenditure that may have responsible for the increased loss of adipose tissue. Due to results of the current investigation, we anticipate that improvements in insulin sensitivity derived from the preferential loss of adipose tissue will maximize the anabolic efficiency to nutrition. Based on earlier studies where we have examined the influence of EAA on muscle protein synthesis [21], a two-sample t-test would have 80% power to detect effect sizes as small as 0.566 in 35 volunteers. It is also important to mention that the individuals in the current study were still deficient in terms of the their total protein intake in the context of caloric restriction, and we are currently revising our formulation to address this issue. In conclusion, we propose that high quality protein+essential amino acids represent a critical variable in the preservation of lean tissue and augmentation of adipose tissue reduction in the elderly population.

REFERENCES

1. Paddon-Jones D, Short KR, Campbell WW, Volpi E, Wolfe RR: Role of dietary protein in the sarcopenia of aging. Am J Clin Nutr 2008, 87(5):1562S-1566S.
2. Hughes VA, Frontera WR, Wood M, Evans WJ, Dallal GE, Roubenoff R, Fiatarone-Singh MA: Longitudinal muscle strength changes in older adults: influence of muscle mass, physical activity, and health. J Geronotol A Biol Sci Med Sci 2001, 56(5):B209-B217.
3. Baumgartner RN, Koehler KM, Gallagher D, et al.: Epidemiology of sarcopenia among the elderly in New Mexico [published erratum appears in. Am J Epidemiol 1999, 149(12):755-763.
4. Honda H, Qureshi AR, Axelsson J, Heimburger O, Suliman ME, Barany P, Stenvinkel P, Lindholm B: Obese sarcopenia in patients with end-stage renal disease is associated with inflammation and increased mortality. Am J Clin Nutr 2007, 86(3):633-638.
5. Janssen I, Heymsfield SB, Ross R: Low relative skeletal muscle mass (sarcopenia) in older persons is associated with functional impairment and physical disability. J Am Geriatr Soc 2002, 50(5):889-896.
6. Cuthbertson D, Smith K, Babraj J, Leese G, Waddell T, Atherton P, Wackerhage H, Taylor PM, Rennie MJ: Anabolic signaling deficits underlie amino acid resistance of wasting, aging muscle. FASEB J 2005, 19(3):422-424.
7. Volpi E, Mittendorfer B, Rasmussen BB, Wolfe RR: The response of muscle protein anabolism to combined hyperaminoacidemia and glucose-induced hyperinsulinemia is impaired in the elderly. J Clin Endocrinol Metab 2000, 85(12):4481-4490.

8. Yang Z, Hall AG: The financial burden of overweight and obesity among elderly Americans: the dynamics of weight, longevity, and health care cost. Health Serv Res 2008, 43(3):849-868.

9. Zamboni M, Mazzali G, Zoico E, Harris TB, Meigs JB, Di Francesco V, Fantin F, Bissoli L, Bosello O: Health consequences of obesity in the elderly: a review of four unresolved questions. Int J Obes (London) 2005, 29(9):1011-1029.

10. Dominguez LJ, Barbagallo M: The cardiometabolic syndrome and sarcopenic obesity in older persons. J Cardiometab Syndr 2007, 2(3):183-189.

11. Miller SL, Wolfe RR: The danger of weight loss in the elderly. J Nutr Health Aging 2008, 12(7):487-491.

12. Campbell WW, Trappe TA, Wolfe RR, Evans WJ: The recommended dietary allowance for protein may not be adequate for older people to maintain skeletal muscle. J Geronotol A Biol Sci Med Sci. 2001, 56(273):M373-M380.

13. Paddon-Jones D, Sheffield-Moore M, Katsanos CS, Zhang XJ, Wolfe RR: Differential stimulation of muscle protein synthesis in elderly humans following isocaloric ingestion of amino acids or whey protein. Exp Gerontol 2006, 41:215-219.

14. Volpi E, Mittendorfer B, Wolf SE, Wolfe RR: Oral amino acids stimulate muscle protein anabolism in the elderly despite higher first-pass splanchnic extraction. Am J Physiol 1999, 277(3 Pt 1):E513-E520.

15. Horton ES: Effects of lifestyle changes to reduce risks of diabetes and associated cardiovascular risks: results from large scale efficacy trial. Obesity (Silver Spring) 2009, 17(Suppl 3):S43-S48.

16. Boersheim E, Bui QU, Tissier S, Kobayashi H, Ferrando AA, Wolfe RR: Effect of amino acid supplementation on muscle mass, strength and physical function in elderly. Clin Nutr 2008, 27(2):189-195.

17. Wolfe RR: Radioactive and stable isotope tracers in biomedicine: principles and practice of kinetic analysis. New York: Wiley-Liss, Inc.; 1992.

18. Volpi E, Kobayashi H, Sheffield-Moore M, Mittendorfer B, Wolfe RR: Essential amino acids are primarily responsible for the amino acid stimulation of muscle protein anabolism in healthy elderly adults. Am J Clin Nutr 2003, 78(2):250-258.

19. Hall KD: Computational model of in vivo human energy metabolism during semistarvation and refeeding. Am J Physiol Endocrinol Metab 2006, 291:E23-E37.

20. Walters SJ: Sample size and power estimation for studies with health related quality of life outcomes: a comparison of four methods using the SF-36. Health Qual Life Outcomes 2004, 25:2-26.

21. Ferrando AA, Paddon-Jones D, Hays NP, Kortebein P, Ronsen O, Williams RH, McComb A, Symons TB, Wolfe RR, Evans WJ: EAA supplementation to increase nitrogen intake improves muscle function during bedrest in the elderly. Clin Nutr 2010, 29(1):18-23.

22. Katsanos CS, Chinkes DL, Paddon-Jones D, Zhang XJ, Aarsland A, Wolfe R: Whey protein ingestion in elderly persons results in greater muscle protein accrual than ingestion of its constituent essential amino acid content. Nutr Res 2008, 28(10):651-658.

23. Schoeller DA, Ravussin E, Schutz Y, Acheson KJ, Baertschi P, Jequier E: Energy expenditure by doubly-labeled water: water validation in humans and proposed calculations. Am J Physiol Endocrinol Metab 1986, 250:R823-R830.

24. Dietary guidelines for Americans. 6th edition. Washington, DC: US Department of Health and Human Services; 2005.
25. Campbell WW, Trappe TA, Wolfe RR, Evans WJ: The recommended dietary allowance for protein may not be adequate for older people to maintain skeletal muscle. J Gerontol A Biol Sci Med Sci 2001, 56:M373-M380.
26. Wolfe RR: The underappreciated role of muscle in health and disease. Am J Clin Nutr 2006, 84(3):475-482.
27. Dietary reference intakes for energy, carbohydrate, fiber, fat, fatty acids, cholesterol, protein, and amino acids (macronutrients). Protein and amino acids. Institute of Medicine, Food and Nutrition Board; Internet: http://www.nap.edu/books/0309085373/html/2002 webcite (accessed 19 June 2006)
28. Han TS, Tajar A, Lean ME: Obesity and weight management in the elderly. Br Med Bull 2011, 97:169-196.
29. Coker RH, Williams RH, Yeo SE, Kortebein PM, Bodenner DL, Kern PA, Evans WJ: The impact of exercise training compared to caloric restriction on hepatic and peripheral insulin resistance in obesity. J Clin Endocrinol Metab 2009, 94(11):4258-4266.
30. Zhao G, Ford ES, Li C, Balluz LS: Physical activity in US older adults with diabetes mellitus: prevalence and correlates of meeting physical activity recommendations. J Am Geriatr Soc 2011, 59(1):132-137.
31. Weiss EC, Galuska DA, Khan LK, Serdula MK: Weight control practices among U.S. adults, 2001–2002. Am J Prev Med 2006, 31:18-24.
32. Tucker JM, Welk GJ, Beyler NK: Physical activity in US Adults compliance with the physical activity guidelines for Americans. Am J Prev Med 2011, 40(4):454-461.
33. Visser M: Obesity, sarcopenia and their functional consequences in old age. Proc Nutr Soc 2011, 70(1):114-118.

Insulinotropic and Muscle Protein Synthetic Effects of Branched-Chain Amino Acids: Potential Therapy for Type 2 Diabetes and Sarcopenia

RALPH J. MANDERS, JONATHAN P. LITTLE, SCOTT C. FORBES, AND DARREN G. CANDOW

5.1 INTRODUCTION

The progressive loss of muscle mass and strength with aging, often referred to as sarcopenia [1] not only decreases overall health, but also increases the dependency on others during activities of daily life and, as such, reduces overall quality of life. Sarcopenia is a multifactorial process characterized by changes in muscle fiber morphology, muscle contractile and protein kinetics, and insulin sensitivity (for reviews see [2,3,4,5]). One main contributing factor towards the age-related loss in muscle mass and strength is the reduced ability to increase skeletal muscle protein syn-

Insulinotropic and Muscle Protein Synthetic Effects of Branched-Chain Amino Acids: Potential Therapy for Type 2 Diabetes and Sarcopenia. © Manders RJ, Little JP, Forbes SC, and Candow DG. Nutrients 4,11 (2012). doi:10.3390/nu4111664. Licensed under a Creative Commons Attribution 3.0 Unported License, http://creativecommons.org/licenses/by/3.0/.

thesis in response to feeding, referred to as "anabolic resistance" (for reviews see [4,6]). Interestingly, ingestion of branched-chain amino acids (BCAA), primarily leucine, increases the activation of the mammalian target of rapamycin (mTOR) signaling pathways involved in muscle protein synthesis via insulin-dependent and independent pathways [7]. High doses of leucine may therefore help overcome "anabolic resistance" to feeding and have a favorable effect on muscle protein synthesis and subsequent maintenance of muscle mass with aging. The rates of muscle protein synthesis are relatively maintained in aging adults [8,9,10]. Leucine exhibits strong insulinotropic characteristics [11,12], which can increase amino acid availability for muscle protein synthesis, inhibit muscle protein breakdown resulting in greater net muscle protein balance over time and also enhance glucose disposal to help maintain blood glucose homeostasis. The purpose of this review is to highlight the potential beneficial health effects of BCAA, primarily leucine, on aging muscle metabolism. We will particularly highlight the role of BCAA in insulin resistance and type 2 diabetes, conditions of which sarcopenia may be a major contributing risk factor.

5.2 BRANCHED CHAIN AMINO ACIDS

The branched chain amino acids (leucine, isoleucine, and valine) account for 14%–18% of the total amino acids in skeletal muscle protein [13,14]. It is well known that amino acids, including the BCAA, are required for maintenance of muscle health in older adults [15]. At rest, BCAA, and particularly leucine, have an anabolic effect through enhanced protein synthesis and/or a decreased rate of protein degradation [16,17,18,19], resulting in a positive net muscle protein balance. Infusion of BCAA in humans elevated the phosphorylation and activation of p70S6 kinase and 4E-BP1 in skeletal muscle [20,21]. Both p70S6 kinase and 4E-BP1 are downstream components of the mTOR signaling pathway, which controls RNA translation and protein synthesis, and is recognized as a central node in support of muscle hypertrophy [22,23,24]. Leucine ingestion is involved in the direct phosphorylation and activation of mTOR in skeletal muscle [25,26], further enhancing the protein synthetic response. However, changes in the rates of muscle protein synthesis are relatively transient unless sufficient amounts of essential amino acids are provided [27], either through normal dietary patterns or supplementation. When BCAA were consumed during and following an acute bout of knee extensor resistance exercise, Karlsson and colleagues [28] found an enhanced (3.5-fold) elevation in p70S6K phosphorylation during recovery in young healthy men compared to resistance exercise alone. The acute exercise-induced p70S6k activity has been shown to correlate with skeletal muscle hypertrophy following 6 weeks of resistance training [29]. In addition, it has been shown that BCAA can attenuate muscle wasting through interaction with the ubiquitin proteasome pathway [30]. This response may partially involve the protein kinase Akt/PKB pathway, which is known to phosphorylate the transcription factor forkhead box class-O (FoxO), that signals downstream to two major ubiquitin ligases atrogin-1 and muscle RING-finger protein (MuRF-1) involved in muscle atrophy [31,32,33].

Aging is known to suppress muscle protein synthesis, especially the synthetic response after feeding, which may alter net muscle protein balance leading to sarcopenia. Although sarcopenia is a multi-factorial affliction [3], amino acids and especially leucine, could play a major role in attenuating the age-related loss in muscle mass and strength. Splanchnic sequestration of leucine following feeding is 50% higher in older vs. younger adults and the rates of muscle protein synthesis are decreased with aging [34], termed "anabolic resistance" [35,36]. Therefore, older adults may require additional dietary protein with greater leucine concentration to counteract muscle wasting over time. Supplemental leucine ingestion has been shown to overcome resistance to the anabolic effects of amino acid consumption [37], providing evidence that leucine supplementation may be beneficial for preserving muscle mass with aging [34,37,38]. For example, the combination of leucine (2.5 g) and casein protein (20 g) elevated the rates of muscle protein synthesis for up to 6 h in older men compared to casein protein ingestion alone [39]. The co-ingestion of leucine (10 g/L) and whey protein (60 g/L) following an acute bout of lower body resistance exercise (6 sets of 10 repetition for leg press and leg extension) in eight older men (75 ± 1 years) significantly increased the rates of muscle protein synthesis and whole-body protein balance [40]. Furthermore, Casperson et al. [41] showed that 2 weeks of leucine supplementation (12 g/day) elevated the muscle protein synthetic response (i.e., augmented mTOR/p70S6K signaling) compared to a standardized meal in older adults without having any effect on lean tissue accretion. It is important to note that acute studies examining phosphorylation or insulin availability after resistance exercise and/or amino acid ingestion are primarily used to predict longer-term training outcomes (i.e., skeletal muscle hypertrophy) and as such, there may be a disconnect between these anabolic signals and end-point measures of protein synthesis [42]. Nevertheless, insulinotropic effects of leucine and/or BCAA may help to improve net muscle protein balance by increasing muscle protein synthesis [31,39,40], decreasing muscle protein breakdown [42], or both. This may be particularly important in long-standing T2D patients where insulin levels are chronically low. Further work, particularly longer-term studies, are warranted to determine if BCAA or leucine have the potential to reverse or prevent sarcopenia, enhance muscle function, and raise the overall quality of life for aging adults.

5.3 INSULINOTROPIC PROPERTIES OF AMINO ACIDS

Next to the anabolic properties of BCAA on muscle health, amino acids can also have profound effects on insulin production/secretion, which could further augment the anabolic response and also be used as a modulator of glucose homeostasis.

The insulinotropic properties of amino acids or protein were reported for the first time in the 1960s [43,44], and have since been confirmed in healthy subjects [45] and type 2 diabetes patients [46,47,48]. In a series of studies, Floyd and co-workers [49,50,51,52,53] reported strong insulinotropic responses following the intravenous administration of various free amino acids. A strong synergistic stimulating effect on insulin release was observed when leucine and arginine were infused in combination with glucose [11]. Furthermore, numerous in vitro studies using primary pancreatic islet cells or β-cell lines have reported strong insulinotropic effects for (among others) leucine, isoleucine, arginine, alanine and phenylalanine [52,54,55,56,57,58,59,60,61]. The mechanisms by which these amino acids stimulate insulin secretion tend to be diverse and have not yet been fully elucidated [62]. Figure 1 provides a simplified overview of amino acid induced insulin secretion. In the presence of glucose, amino acids such as arginine have been shown to stimulate insulin secretion by directly depolarizing the plasma membrane of the β-cell [54], which opens up voltage activated Ca^{2+} channels, leading to the influx of Ca^{2+} and subsequent insulin exocytosis [57,62]. Other amino acids may modulate their insulinotropic properties through activating Ca^{2+} channels by their co-transport with Na^+ [57,63]. Furthermore, intracellular catabolism of all metabolizable amino acids will increase the intracellular ATP/ADP ratio, thereby closing ATP-sensitive K^+ channels, which can also lead to the depolarization of the plasma membrane [62,64,65]. Both in vivo and in vitro work has identified leucine as a particular interesting insulin secretagogue as it both induces and enhances pancreatic β-cell insulin secretion through its oxidative decarboxylation, as well as by its ability to allosterically activate glutamate dehydrogenase [60,62,66,67] which increases ATP/ADP ratios by increasing TCA-cycle fluxes resulting in depolarization of the plasma membrane through closure of ATP-sensitive K^+ channels. Furthermore, leucine can be transaminated to α-ketoisocaproate which in turn is converted into acetyl-CoA before entering the TCA-cycle [68]. These findings are in line with recent in vivo observations, showing co-ingestion of relatively small amounts of free leucine to further augment the insulin response following the combined ingestion of carbohydrate and protein in healthy men [69]. Xu et al. [60] reported that the same signals that stimulate insulin release are also likely to be responsible for the leucine-induced activation of the mammalian target of rapamycin (mTOR) signaling pathway in the pancreatic β-cell. The potency of leucine to activate protein synthesis by interacting with the mTOR signaling pathway has been proposed to enhance β-cell function through the maintenance of β-cell mass. As such, the insulinotropic properties of amino acids (and leucine in particular) can therefore be of great clinical relevance in the treatment of type 2 diabetes or any state where there is a certain level of insulin resistance (e.g., aging) or hyperglycemia. Increasing endogenous insulin secretion with amino acids could therefore accelerate blood glucose disposal resulting in a better glycemic control. In longstanding type 2 diabetes patients, hyperglycemia is no longer accompanied by a compensatory hyperinsulinemia and as such, it is generally assumed that the capacity of the β-cell to secrete insulin is severely impaired due to several defects [70]. These defects include a reduced early insulin secretory response to oral glucose, a reduced ability of the β-cell to compensate for the degree of insulin resistance, a decline in the glucose-sensing ability of the β-cell, and a shift to the right in the dose-response curve relating glucose and insulin secretion, which are all indicative of a progressive insensitivity of the β-cell to glucose [71]. All these defects involve glucose-sensing and -signaling pathways in the β-cell. Although insulin secretion in response to carbohydrate intake is impaired in type 2 diabetes patients, it has been shown that the combined ingestion of a protein/amino acid mixture with carbohydrates can increase the plasma insulin response up to 4-fold [72,73]. This indicates that although the sensitivity of the pancreas to carbohydrate intake is significantly reduced in type 2 diabetes patients, the capacity to secrete insulin in response to stimuli like amino acids is still intact. Therefore, it can be concluded that the defects in the insulin response after a meal or glucose load in these patients can mainly be attributed to the reduced sensitivity of the β-cell to glucose, and not an overall defect in the capacity to produce and/or secrete insulin. For this reason it can also be assumed that the potential of amino acids is the greatest in longstanding type 2 diabetes patients as they are no longer in a state of hyperinsulinemia, in contrast to recently diagnosed patients where hyperglycemia is accompanied with hyperinsulinemia. Before amino acid supplementation or modulation can be considered an effective nutritional intervention in the treatment of type 2 diabetes, the mere increase in endogenous insulin production alone is not sufficient. In order to improve glycemic control (i.e., lower blood glucose concentrations), the increased endogenous insulin secretory response should be able to overcome the level of insulin resistance and effectively lower plasma glucose concentrations by increasing the glucose disposal rate from the circulation. Using stable isotope glucose tracers, it has been shown that even though plasma glucose disposal was severely impaired in the type 2 diabetes patients, the addition of an amino acid/protein mixture increased plasma glucose disposal significantly and results in a lower glycemic responses [73]. As mentioned previously, leucine has been identified as a particular interesting insulin secretagogue [60,62,66,67] and in an effort to determine the specific role of leucine supplementation an insulinotropic protein mixture was tested together with a single, meal-like, bolus of carbohydrate in longstanding type 2 diabetes patients. Addition of free leucine to the mixture significantly increased circulating insulin concentrations but failed to result in a further improvement in glycemic control [12]. In a series of real life studies, these results have also been confirmed. In these studies the effects of protein modulation (protein only, and protein combined with free leucine) were determined on the prevalence of hyperglycemia in well-controlled type 2 diabetes patients as measured with continuous glucose monitoring. Whereas a protein/leucine mixture was able to lower the prevalence of hyperglycemia by 26% [74], the same absolute amount of protein (without free leucine) did not results in a further improvement of glycemic control [75]. These results extend on previous findings [46,76,77], and imply that nutritional interventions with protein, and leucine in particular can represent an effective strategy to reduce postprandial blood glucose excursions.

However, no long-term studies have focused on the question of whether the insulinotropic potential of amino acids remains after being on a high protein diet or amino acids supplementation. It therefore remains to be seen whether such a nutritional intervention represents a feasible long-term strategy to improve glycemic control. Next to the insulinotropic properties of amino acids and dietary protein there are several other beneficial effects that could result in a better health status for both diabetics, elderly or obese subjects.

In long-term dietary interventions, protein and leucine supplementation would eventually lead to changes in the macronutrient composition of an ad libitum diet, while keeping the person in energy balance. The greater protein intake would be accompanied by a reduction in total dietary fat and carbohydrate consumption. This kind of dietary modulation should result in an even further improvement in glycemic control as total carbohydrate intake is lower. In accordance, increasing the protein content of the diet, at the expense of carbohydrate and fat, drastically lowered blood glucose concentrations in a group type 2 diabetes patients over a 5 week intervention period [78,79,80]. Furthermore, it should be noted that diets high in protein have been reported to be more effective when trying to maintain body weight after a period of weight loss when compared to high carbohydrate diets. This benefit has been attributed to the thermogenic and satiating properties of dietary protein [81,82,83,84,85], which in the long run can further optimize glycemic control.

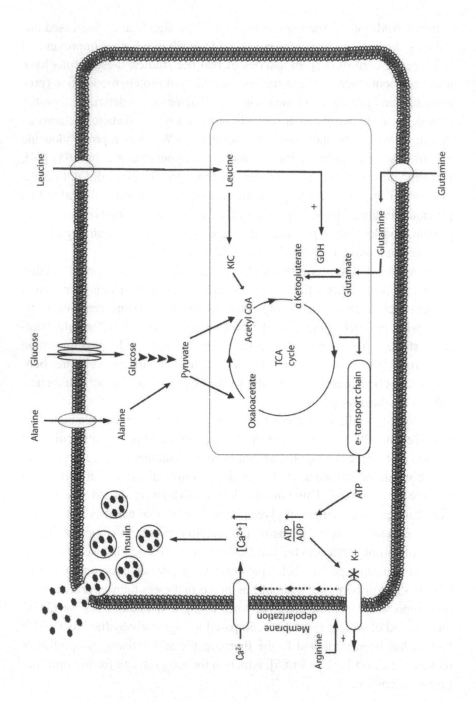

FIGURE 1: Simplified overview of amino acid induced insulin secretion in the β-cell. Glucose is metabolized in the cell via glycolysis into pyruvate, which is subsequently metabolized further by the tricaboxylic acid cycle (TCA cycle) to form ATP. Increased ratios of ATP/ADP result in depolarization of the plasma membrane through closure of ATP-sensitive K^+ channels. This depolarization opens voltage activated Ca^{2+} channels, leading to increased concentrations of intracellular Ca^{2+} ($[Ca^{2+}]_i$) and subsequent insulin exocytosis. Intracellular catabolism of all metabolizable amino acids will increase the intracellular ATP/ADP ratio, thereby closing ATP-sensitive K^+ channels, leading to the depolarization of the plasma membrane. Leucine both induces and enhances pancreatic β-cell insulin secretion through oxidative decarboxylation and allosteric activation of glutamate dehydrogenase (GDH) increasing ATP/ADP. Leucine can also be transaminated to α-ketoisocaproate (KIC) that is converted into acetyl-CoA before entering the TCA-cycle. Amino acids such as arginine can directly depolarize the plasma membrane of the β-cell, opening up voltage activated Ca^{2+} channels leading to insulin secretion. Adapted from Newsholme et al. [86].

Though there are ample suggestions that amino acid or protein supplementation could represent an effective dietary strategy to improve blood glucose homeostasis in type 2 diabetes, future research should determine if these insulinotropic properties are retained after prolonged increase of dietary protein or BCAA supplementation.

5.4 BCAA AND MUSCLE METABOLIC HEALTH

In addition to the potential for BCAA (and leucine in particular) to benefit metabolic health through hypertrophic or insulinotropic pathways (as described in previous sections), there is emerging evidence that BCAA may also influence skeletal muscle metabolism. Because skeletal muscle is responsible for ~75%–80% of glucose disposal in response to carbohydrate ingestion [87] and is a main contributor to metabolic rate, alterations in skeletal muscle metabolism have profound effects on whole body metabolic health. A loss of skeletal muscle mass, for example, reduces overall glucose disposal capacity, which can result in elevated circulating glucose concentrations, unrelated to the level of insulin sensitivity. Mitochondria are organelles responsible for generating cellular energy through oxidation of substrates and are therefore critical to metabolic regulation. Accordingly, reduced quantity and/or quality of skeletal muscle mitochondria is hypothesized to contribute to insulin resistance in older adults [88] and aging-related diseases, such as type 2 diabetes [89,90]. Mitochondria also play important roles in oxidative stress and apoptosis, which are clearly implicated in the aging process. As such, strategies that can increase, or preserve, muscle mitochondrial function may have therapeutic benefit in aging.

Recently, D'Antona and colleagues [91] fed mice a diet enriched in BCAA and demonstrated that average lifespan was increased. These authors linked the anti-aging effects of BCAA supplementation to increased mitochondrial biogenesis in skeletal muscle and heart, and demonstrated that markers of oxidative stress were reduced. Earlier findings had indicated that dietary leucine supplementation could improve glucose regulation in mice with diet-induced obesity [92]. These beneficial metabolic effects of BCAA supplementation in rodents are supported by some human data, where 60 weeks of AA supplementation (containing relatively high proportion of BCAA) was shown to improve insulin sensitivity and glucose control in a small trial involving elderly patients with type 2 diabetes [93]. Thus, it is possible that BCAA supplementation could have benefits to metabolic health through mechanisms that improve skeletal muscle mitochondria mass and function [94].

These potential positive effects of BCAA supplementation for metabolic health and aging must be balanced against any negative outcomes. In this regard, recent human studies have found potential links between elevated plasma BCAA and obesity/type 2 diabetes. Using metabolomics profiling, Newgard and colleagues [95] demonstrated that BCAA were elevated in obese humans, and suggested that BCAA overload may contribute to insulin resistance. In a separate study, circulating BCAA (along with phenylalanine and tyrosine) demonstrated high associations with the development of type 2 diabetes in a group of 2422 individuals who had blood samples taken at baseline and were followed for 12 years [96]. It has been proposed that elevated BCAA concentrations result in an over-activation of mTOR/p70S6 kinase which, in turn, results in an increased IRS-1 phosphorylation on serine residues thereby inhibiting PI3 kinase

[94]. This inhibition of PI3K in turn leads to impaired insulin signaling and contributes to insulin resistance [97,98,99]. Future studies are needed to decipher whether these findings are (i) indicative of BCAA contributing to impaired metabolic health, (ii) the result of impaired BCAA catabolism in obesity/diabetes, or (iii) are a possible compensatory mechanism in response to obesity.

5.5 CONCLUSIONS

With increasing prevalence of both sarcopenia and type 2 diabetes, there is a great need for novel interventions that effectively combat the loss of skeletal muscle mass and increased insulin resistance that play a pivotal role in both afflictions. Amino acids in general, and the branched-chain amino acids in particular, are likely to represent a nutritional approach that is able to reduce or revert the age-related loss of muscle mass and function, overcome anabolic resistance and improve glycemic control. Leucine consumption increases the activation of the mTOR signaling pathways involved in muscle protein synthesis through insulin-dependent and independent pathways and may therefore overcome the anabolic resistance to nutrition and help to maintain muscle mass in an aging population. Apart from their anabolic properties, amino acids also exhibit strong insulinotropic effects as they can either directly induce insulin secretion or function as a substrate to increase ATP/ADP ratio's that stimulate insulin exocytosis. The insulinotropic effects of BCAA may exert further influence on positive muscle protein balance by reducing muscle protein breakdown. Furthermore, the same signaling pathways that lead to skeletal muscle protein synthesis also play a role in enhancing β-cell mass and function. BCAA can also influence skeletal muscle metabolism by improving the quantity and quality skeletal muscle mitochondria, and as such, increase, or preserve, muscle mitochondrial function that may have therapeutic benefits in aging. However, more long-term studies are warranted to fully elucidate the true potential of the anabolic and insulinotropic potential of amino acids.

REFERENCES

1. Cruz-Jentoft, A.J.; Baeyens, J.P.; Bauer, J.M.; Boirie, Y.; Cederholm, T.; Landi, F.; Martin, F.C.; Michel, J.P.; Rolland, Y.; Schneider, S.M.; et al. Sarcopenia: European consensus on definition and diagnosis: Report of the European Working Group on Sarcopenia in Older People. Age Ageing 2010, 39, 412–423.
2. Candow, D.G.; Forbes, S.C.; Little, J.P.; Cornish, S.M.; Pinkoski, C.; Chilibeck, P.D. Effect of nutritional interventions and resistance exercise on aging muscle mass and strength. Biogerontology 2012, 13, 345–358.
3. Forbes, S.C.; Little, J.P.; Candow, D.G. Exercise and nutritional interventions for improving aging muscle health. Endocrine 2012, 42, 29–38.
4. Haran, P.H.; Rivas, D.A.; Fielding, R.A. Role and potential mechanisms of anabolic resistance in sarcopenia. J. Cachexia Sarcopenia Muscle 2012, 3, 157–162.
5. Leenders, M.; van Loon, L.J. Leucine as a pharmaconutrient to prevent and treat sarcopenia and type 2 diabetes. Nutr. Rev. 2011, 69, 675–689.
6. Breen, L.; Phillips, S.M. Skeletal muscle protein metabolism in the elderly: Interventions to counteract the "anabolic resistance" of ageing. Nutr. Metab. 2011, 8, 68.
7. Kimball, S.R.; Jefferson, L.S. Regulation of global and specific mRNA translation by oral administration of branched-chain amino acids. Biochem. Biophys. Res. Commun. 2004, 313, 423–427.
8. Cuthbertson, D.; Smith, K.; Babraj, J.; Leese, G.; Waddell, T.; Atherton, P.; Wackerhage, H.; Taylor, P.M.; Rennie, M.J. Anabolic signaling deficits underlie amino acid resistance of wasting, aging muscle. FASEB J. 2005, 19, 422–424.
9. Pereira, S.; Marliss, E.B.; Morais, J.; Chevalier, S.; Gougeon, R. Insulin resistance of protein metabolism in type 2 diabetes mellitus. Diabetes 2008, 57, 56–63.
10. Welle, S.; Thornton, C.; Statt, M.; McHenry, B. Postprandial myofibrillar and whole body protein synthesis in young and old human subjects. Am. J. Physiol. 1994, 267, E599–E604.
11. Floyd, J.C.; Fajans, S.S.; Pek, S.; Thiffault, C.A.; Knopf, R.F.; Conn, J.W. Synergistic effect of essential amino acids and glucose upon insulin secretion in man. Diabetes 1970, 19, 109–115.
12. Manders, R.J.; Koopman, R.; Sluijsmans, W.E.; van den Berg, R.; Verbeek, K.; Saris, W.H.; Wagenmakers, A.J.; van Loon, L.J. Co-ingestion of a protein hydrolysate with or without additional leucine effectively reduces post-prandial blood glucose excursions in Type 2 diabetic men. J. Nutr. 2006, 136, 1294–1299.
13. Riazi, R.; Wykes, L.J.; Ball, R.O.; Pencharz, P.B. The total branched-chain amino acid requirement in young healthy adult men determined by indicator amino acid oxidation by use of L-[1-13C]phenylalanine. J. Nutr. 2003, 133, 1383–1389.
14. Layman, D.K.; Baum, J.I. Dietary protein impact on glycemic control during weight loss. J. Nutr. 2004, 134, 968S–973S.
15. Millward, D.J. Sufficient protein for our elders? Am. J. Clin. Nutr. 2008, 88, 1187–1188.
16. Kimball, S.R.; Jefferson, L.S. Signaling pathways and molecular mechanisms through which branched-chain amino acids mediate translational control of protein synthesis. J. Nutr. 2006, 136, 227S–231S.

17. Alvestrand, A.; Hagenfeldt, L.; Merli, M.; Oureshi, A.; Eriksson, L.S. Influence of leucine infusion on intracellular amino acids in humans. Eur. J. Clin. Invest. 1990, 20, 293–298.

18. Louard, R.J.; Barrett, E.J.; Gelfand, R.A. Effect of infused branched-chain amino acids on muscle and whole-body amino acid metabolism in man. Clin. Sci. (Lond.) 1990, 79, 457–466.

19. Nair, K.S.; Schwartz, R.G.; Welle, S. Leucine as a regulator of whole body and skeletal muscle protein metabolism in humans. Am. J. Physiol. 1992, 263, E928–E934.

20. Greiwe, J.S.; Kwon, G.; McDaniel, M.L.; Semenkovich, C.F. Leucine and insulin activate p70 S6 kinase through different pathways in human skeletal muscle. Am. J. Physiol. Endocrinol. Metab. 2001, 281, E466–E471.

21. Liu, Z.; Long, W.; Fryburg, D.A.; Barrett, E.J. The regulation of body and skeletal muscle protein metabolism by hormones and amino acids. J. Nutr. 2006, 136, 212S–217S.

22. Kimball, S.R. The role of nutrition in stimulating muscle protein accretion at the molecular level. Biochem. Soc. Trans. 2007, 35, 1298–1301.

23. Kimball, S.R.; Jefferson, L.S. Molecular mechanisms through which amino acids mediate signaling through the mammalian target of rapamycin. Curr. Opin. Clin. Nutr. Metab. Care 2004, 7, 39–44.

24. Rennie, M.J.; Wackerhage, H.; Spangenburg, E.E.; Booth, F.W. Control of the size of the human muscle mass. Annu. Rev. Physiol. 2004, 66, 799–828.

25. Fujita, S.; Dreyer, H.C.; Drummond, M.J.; Glynn, E.L.; Cadenas, J.G.; Yoshizawa, F.; Volpi, E.; Rasmussen, B.B. Nutrient signalling in the regulation of human muscle protein synthesis. J. Physiol. 2007, 582, 813–823.

26. Deldicque, L.; Theisen, D.; Francaux, M. Regulation of mTOR by amino acids and resistance exercise in skeletal muscle. Eur. J. Appl. Physiol. 2005, 94, 1–10.

27. Kobayashi, H.; Kato, H.; Hirabayashi, Y.; Murakami, H.; Suzuki, H. Modulations of muscle protein metabolism by branched-chain amino acids in normal and muscle-atrophying rats. J. Nutr. 2006, 136, 234S–236S.

28. Karlsson, H.K.; Nilsson, P.A.; Nilsson, J.; Chibalin, A.V.; Zierath, J.R.; Blomstrand, E. Branched-chain amino acids increase p70S6k phosphorylation in human skeletal muscle after resistance exercise. Am. J. Physiol. Endocrinol. Metab. 2004, 287, E1–E7.

29. Baar, K.; Esser, K. Phosphorylation of p70(S6k) correlates with increased skeletal muscle mass following resistance exercise. Am. J. Physiol. Endocrinol. Metab. 1999, 276, C120–C127.

30. Busquets, S.; Alvarez, B.; Llovera, M.; Agell, N.; Lopez-Soriano, F.J.; Argiles, J.M. Branched-chain amino acids inhibit proteolysis in rat skeletal muscle: Mechanisms involved. J. Cell. Physiol. 2000, 184, 380–384.

31. Nicastro, H.; Artioli, G.G.; Costa Ados, S.; Solis, M.Y.; da Luz, C.R.; Blachier, F.; Lancha, A.H., Jr. An overview of the therapeutic effects of leucine supplementation on skeletal muscle under atrophic conditions. Amino Acids 2011, 40, 287–300.

32. Sandri, M. Signaling in muscle atrophy and hypertrophy. Physiology 2008, 23, 160–170.

33. Glass, D.J. Skeletal muscle hypertrophy and atrophy signaling pathways. Inter. J. Biochem. Cell Biol. 2005, 37, 1974–1984.

34. Fujita, S.; Volpi, E. Amino acids and muscle loss with aging. J. Nutr. 2006, 136, 277S–280S.
35. Burd, N.A.; Wall, B.T.; van Loon, L.J. The curious case of anabolic resistance: Old wives' tales or new fables? J. Appl. Physiol. 2012, 112, 1233–1235.
36. Rennie, M.J. Anabolic resistance: The effects of aging, sexual dimorphism, and immobilization on human muscle protein turnover. Appl. Physiol. Nutr. Metab. 2009, 34, 377–381.
37. Katsanos, C.S.; Kobayashi, H.; Sheffield-Moore, M.; Aarsland, A.; Wolfe, R.R. A high proportion of leucine is required for optimal stimulation of the rate of muscle protein synthesis by essential amino acids in the elderly. Am. J. Physiol. Endocrinol. Metab. 2006, 291, E381–E387.
38. Dreyer, H.C.; Drummond, M.J.; Pennings, B.; Fujita, S.; Glynn, E.L.; Chinkes, D.L.; Dhanani, S.; Volpi, E.; Rasmussen, B.B. Leucine-enriched essential amino acid and carbohydrate ingestion following resistance exercise enhances mTOR signaling and protein synthesis in human muscle. Am. J. Physiol. Endocrinol. Metab. 2008, 294, E392–E400.
39. Wall, B.T.; Hamer, H.M.; de Lange, A.; Kiskini, A.; Groen, B.B.; Senden, J.M.; Gijsen, A.P.; Verdijk, L.B.; van Loon, L.J. Leucine co-ingestion improves post-prandial muscle protein accretion in elderly men. Clin. Nutr. 2012.
40. Koopman, R.; Verdijk, L.; Manders, R.J.; Gijsen, A.P.; Gorselink, M.; Pijpers, E.; Wagenmakers, A.J.; van Loon, L.J. Co-ingestion of protein and leucine stimulates muscle protein synthesis rates to the same extent in young and elderly lean men. Am. J. Clin. Nutr. 2006, 84, 623–632.
41. Casperson, S.L.; Sheffield-Moore, M.; Hewlings, S.J.; Paddon-Jones, D. Leucine supplementation chronically improves muscle protein synthesis in older adults consuming the RDA for protein. Clin. Nutr. 2012, 31, 512–519.
42. Greenhaff, P.L.; Karagounis, L.G.; Peirce, N.; Simpson, E.J.; Hazell, M.; Layfield, R.; Wackerhage, H.; Smith, K.; Atherton, P.; Selby, A.; Rennie, M.J. Disassociation between the effects of amino acids and insulin on signaling, ubiquitin ligases, and protein turnover in human muscle. Am. J. Physiol. Endocrinol. Metab. 2008, 295, E595–E604.
43. Pallotta, J.A.; Kennedy, P.J. Response of plasma insulin and growth hormone to carbohydrate and protein feeding. Metabolism 1968, 17, 901–908.
44. Rabinowitz, D.; Merimee, T.J.; Maffezzoli, R.; Burgess, J.A. Patterns of hormonal release after glucose, protein, and glucose plus protein. Lancet 1966, 2, 454–456.
45. Nuttall, F.Q.; Gannon, M.C.; Wald, J.L.; Ahmed, M. Plasma glucose and insulin profiles in normal subjects ingesting diets of varying carbohydrate, fat, and protein content. J. Am. Coll. Nutr. 1985, 4, 437–450.
46. Nuttall, F.Q.; Mooradian, A.D.; Gannon, M.C.; Billington, C.; Krezowski, P. Effect of protein ingestion on the glucose and insulin response to a standardized oral glucose load. Diabetes Care 1984, 7, 465–470.
47. Gannon, M.C.; Nuttall, F.Q.; Lane, J.T.; Burmeister, L.A. Metabolic response to cottage cheese or egg white protein, with or without glucose, in type II diabetic subjects. Metabolism 1992, 41, 1137–1145.

48. Gannon, M.C.; Nuttall, F.Q.; Grant, C.T.; Ercan-Fang, S.; Ercan-Fang, N. Stimulation of insulin secretion by fructose ingested with protein in people with untreated type 2 diabetes. Diabetes Care 1998, 21, 16–22.
49. Fajans, S.S.; Knopf, R.F.; Floyd, J.C., Jr.; Power, L.; Conn, J.W. The experimental induction in man of sensitivity to leucine hypoglycemia. J. Clin. Invest. 1963, 42, 216–229.
50. Floyd, J.C., Jr.; Fajans, S.S.; Conn, J.W.; Knopf, R.F.; Rull, J. Stimulation of insulin secretion by amino acids. J. Clin. Invest. 1966, 45, 1487–1502.
51. Floyd, J.C., Jr.; Fajans, S.S.; Conn, J.W.; Thiffault, C.; Knopf, R.F.; Guntsche, E. Secretion of insulin induced by amino acids and glucose in diabetes mellitus. J. Clin. Endocrinol. Metab. 1968, 28, 266–276.
52. Floyd, J.C., Jr.; Fajans, S.S.; Knopf, R.F.; Conn, J.W. Evidence that insulin release is the mechanism for experimentally induced leucine hypoglycemia in man. J. Clin. Invest. 1963, 42, 1714–1719.
53. Floyd, J.C., Jr.; Fajans, S.S.; Pek, S.; Thiffault, C.A.; Knopf, R.F.; Conn, J.W. Synergistic effect of certain amino acid pairs upon insulin secretion in man. Diabetes 1970, 19, 102–108.
54. Blachier, F.; Leclercq-Meyer, V.; Marchand, J.; Woussen Colle, M.C.; Mathias, P.C.; Sener, A.; Malaisse, W.J. Stimulus-secretion coupling of arginine-induced insulin release. Functional response of islets to L-arginine and L-ornithine. Biochim. Biophys. Acta 1989, 1013, 144–151.
55. Sener, A.; Hutton, J.C.; Malaisse, W.J. The stimulus-secretion coupling of amino acid-induced insulin release. Synergistic effects of L-glutamine and 2-keto acids upon insulin secretion. Biochim. Biophys. Acta 1981, 677, 32–38.
56. Malaisse, W.J.; Plasman, P.O.; Blachier, F.; Herchuelz, A.; Sener, A. Stimulus-secretion coupling of arginine-induced insulin release: Significance of changes in extracellular and intracellular pH. Cell Biochem. Funct. 1991, 9, 1–7.
57. McClenaghan, N.H.; Barnett, C.R.; O'Harte, F.P.; Flatt, P.R. Mechanisms of amino acid-induced insulin secretion from the glucose-responsive BRIN-BD11 pancreatic B-cell line. J. Endocrinol. 1996, 151, 349–357.
58. Pipeleers, D.G.; Schuit, F.C.; in't Veld, P.A.; Maes, E.; Hooghe-Peters, E.L.; van de Winkel, M.; Gepts, W. Interplay of nutrients and hormones in the regulation of insulin release. Endocrinology 1985, 117, 824–833.
59. Schwanstecher, C.; Meyer, M.; Schwanstecher, M.; Panten, U. Interaction of N-benzoyl-D-phenylalanine and related compounds with the sulphonylurea receptor of beta-cells. Br. J. Pharmacol. 1998, 123, 1023–1030.
60. Xu, G.; Kwon, G.; Cruz, W.S.; Marshall, C.A.; McDaniel, M.L. Metabolic regulation by leucine of translation initiation through the mTOR-signaling pathway by pancreatic beta-cells. Diabetes 2001, 50, 353–360.
61. Lajoix, A.D.; Reggio, H.; Chardes, T.; Peraldi-Roux, S.; Tribillac, F.; Roye, M.; Dietz, S.; Broca, C.; Manteghetti, M.; Ribes, G.; Wollheim, C.B.; Gross, R. A neuronal isoform of nitric oxide synthase expressed in pancreatic beta-cells controls insulin secretion. Diabetes 2001, 50, 1311–1323.

62. Newsholme, P.; Brennan, L.; Rubi, B.; Maechler, P. New insights into amino acid metabolism, beta-cell function and diabetes. Clin. Sci. (Lond.) 2005, 108, 185–194.
63. Sener, A.; Malaisse, W.J. The stimulus-secretion coupling of amino acid-induced insulin release. Insulinotropic action of L-alanine. Biochim. Biophys. Acta 2002, 1573, 100–104.
64. Dunne, M.J.; Yule, D.I.; Gallacher, D.V.; Petersen, O.H. Effects of alanine on insulin-secreting cells: Patch-clamp and single cell intracellular Ca2+ measurements. Biochim. Biophys. Acta 1990, 1055, 157–164.
65. Brennan, L.; Shine, A.; Hewage, C.; Malthouse, J.P.; Brindle, K.M.; McClenaghan, N.; Flatt, P.R.; Newsholme, P. A nuclear magnetic resonance-based demonstration of substantial oxidative L-alanine metabolism and L-alanine-enhanced glucose metabolism in a clonal pancreatic beta-cell line: Metabolism of L-alanine is important to the regulation of insulin secretion. Diabetes 2002, 51, 1714–1721.
66. Fahien, L.A.; MacDonald, M.J.; Kmiotek, E.H.; Mertz, R.J.; Fahien, C.M. Regulation of insulin release by factors that also modify glutamate dehydrogenase. J. Biol. Chem. 1988, 263, 13610–13614.
67. Sener, A.; Malaisse, W.J. L-leucine and a nonmetabolized analogue activate pancreatic islet glutamate dehydrogenase. Nature 1980, 288, 187–189.
68. Panten, U.; Kriegstein, E.; Poser, W.; Schonborn, J.; Hasselblatt, A. Effects of L-leucine and alpha-ketoisocaproic acid upon insulin secretion and metabolism of isolated pancreatic islets. FEBS Lett. 1972, 20, 225–228.
69. Koopman, R.; Wagenmakers, A.J.; Manders, R.J.; Zorenc, A.H.; Senden, J.M.; Gorselink, M.; Keizer, H.A.; van Loon, L.J. Combined ingestion of protein and free leucine with carbohydrate increases postexercise muscle protein synthesis in vivo in male subjects. Am. J. Physiol. Endocrinol. Metab. 2005, 288, E645–E653.
70. Porte, D., Jr.; Kahn, S.E. beta-cell dysfunction and failure in type 2 diabetes: Potential mechanisms. Diabetes 2001, 50, S160–S163.
71. Polonsky, K.S.; Sturis, J.; Bell, G.I. Seminars in Medicine of the Beth Israel Hospital, Boston. Non-insulin-dependent diabetes mellitus—a genetically programmed failure of the beta cell to compensate for insulin resistance. N. Engl. J. Med. 1996, 334, 777–783.
72. Van Loon, L.J.C.; Kruishoop, M.; Menheere, P.P.C.A.; Wagenmakers, A.J.M.; Saris, W.H.M.; Keizer, H.A. Amino acid ingestion strongly enhances insulin secretion in patients with long-term type 2 diabetes. Diabetes Care 2003, 26, 625–630.
73. Manders, R.J.; Wagenmakers, A.J.; Koopman, R.; Zorenc, A.H.; Menheere, P.P.; Schaper, N.C.; Saris, W.H.; van Loon, L.J. Co-ingestion of a protein hydrolysate and amino acid mixture with carbohydrate improves plasma glucose disposal in patients with type 2 diabetes. Am. J. Clin. Nutr. 2005, 82, 76–83.
74. Manders, R.J.; Praet, S.F.; Meex, R.C.; Koopman, R.; de Roos, A.L.; Wagenmakers, A.J.; Saris, W.H.; van Loon, L.J. Protein hydrolysate/leucine co-ingestion reduces the prevalence of hyperglycemia in type 2 diabetic patients. Diabetes Care 2006, 29, 2721–2722.
75. Manders, R.J.; Praet, S.F.; Vikstrom, M.H.; Saris, W.H.; van Loon, L.J. Protein hydrolysate co-ingestion does not modulate 24 h glycemic control in long-standing type 2 diabetes patients. Eur. J. Clin. Nutr. 2009, 63, 121–126.

76. Frid, A.H.; Nilsson, M.; Holst, J.J.; Bjorck, I.M. Effect of whey on blood glucose and insulin responses to composite breakfast and lunch meals in type 2 diabetic subjects. Am. J. Clin. Nutr. 2005, 82, 69–75.

77. Nilsson, M.; Stenberg, M.; Frid, A.H.; Holst, J.J.; Bjorck, I.M. Glycemia and insulinemia in healthy subjects after lactose-equivalent meals of milk and other food proteins: The role of plasma amino acids and incretins. Am. J. Clin. Nutr. 2004, 80, 1246–1253.

78. Gannon, M.C.; Nuttall, F.Q. Effect of a high-protein, low-carbohydrate diet on blood glucose control in people with type 2 diabetes. Diabetes 2004, 53, 2375–2382.

79. Gannon, M.C.; Nuttall, F.Q.; Saeed, A.; Jordan, K.; Hoover, H. An increase in dietary protein improves the blood glucose response in persons with type 2 diabetes. Am. J. Clin. Nutr. 2003, 78, 734–741.

80. Nuttall, F.Q.; Schweim, K.; Hoover, H.; Gannon, M.C. Effect of the LoBAG30 diet on blood glucose control in people with type 2 diabetes. Br. J. Nutr. 2008, 99, 511–519.

81. Astrup, A. The satiating power of protein—a key to obesity prevention? Am. J. Clin. Nutr. 2005, 82, 1–2.

82. Weigle, D.S.; Breen, P.A.; Matthys, C.C.; Callahan, H.S.; Meeuws, K.E.; Burden, V.R.; Purnell, J.Q. A high-protein diet induces sustained reductions in appetite, ad libitum caloric intake, and body weight despite compensatory changes in diurnal plasma leptin and ghrelin concentrations. Am. J. Clin. Nutr. 2005, 82, 41–48.

83. Lejeune, M.P.; Kovacs, E.M.; Westerterp-Plantenga, M.S. Additional protein intake limits weight regain after weight loss in humans. Br. J. Nutr. 2005, 93, 281–289.

84. Westerterp-Plantenga, M.S.; Lejeune, M.P. Protein intake and body-weight regulation. Appetite 2005, 45, 187–190.

85. Due, A.; Toubro, S.; Skov, A.R.; Astrup, A. Effect of normal-fat diets, either medium or high in protein, on body weight in overweight subjects: A randomised 1-year trial. Int. J. Obes. Relat. Metab. Disord. 2004, 28, 1283–1290.

86. Newsholme, P.; Brennan, L.; Bender, K. Amino Acid Metabolism, {beta}-Cell Function, and Diabetes. Diabetes 2006, 55, S39–S47.

87. Thiebaud, D.; Jacot, E.; DeFronzo, R.A.; Maeder, E.; Jequier, E.; Felber, J.P. The effect of graded doses of insulin on total glucose uptake, glucose oxidation, and glucose storage in man. Diabetes 1982, 31, 957–963.

88. Petersen, K.F.; Befroy, D.; Dufour, S.; Dziura, J.; Ariyan, C.; Rothman, D.L.; DiPietro, L.; Cline, G.W.; Shulman, G.I. Mitochondrial dysfunction in the elderly: Possible role in insulin resistance. Science 2003, 300, 1140–1142.

89. Petersen, K.F.; Dufour, S.; Befroy, D.; Garcia, R.; Shulman, G.I. Impaired Mitochondrial Activity in the Insulin-Resistant Offspring of Patients with Type 2 Diabetes. N. Engl. J. Med. 2004, 350, 664–671.

90. Mogensen, M.; Sahlin, K.; Fernstrom, M.; Glintborg, D.; Vind, B.F.; Beck-Nielsen, H.; Hojlund, K. Mitochondrial Respiration Is Decreased in Skeletal Muscle of Patients With Type 2 Diabetes. Diabetes 2007, 56, 1592–1599.

91. D'Antona, G.; Ragni, M.; Cardile, A.; Tedesco, L.; Dossena, M.; Bruttini, F.; Caliaro, F.; Corsetti, G.; Bottinelli, R.; Carruba, M.O.; Valerio, A.; Nisoli, E. Branched-Chain Amino Acid Supplementation Promotes Survival and Supports Cardiac and

Skeletal Muscle Mitochondrial Biogenesis in Middle-Aged Mice. Cell Metab. 2010, 12, 362–372.

92. Zhang, Y.; Guo, K.; LeBlanc, R.E.; Loh, D.; Schwartz, G.J.; Yu, Y.H. Increasing dietary leucine intake reduces diet-induced obesity and improves glucose and cholesterol metabolism in mice via multimechanisms. Diabetes 2007, 56, 1647–1654.

93. Solerte, S.B.; Fioravanti, M.; Locatelli, E.; Bonacasa, R.; Zamboni, M.; Basso, C.; Mazzoleni, A.; Mansi, V.; Geroutis, N.; Gazzaruso, C. Improvement of Blood Glucose Control and Insulin Sensitivity during a Long-Term (60 Weeks) Randomized Study with Amino Acid Dietary Supplements in Elderly Subjects with Type 2 Diabetes Mellitus. Am. J. Cardiol. 2008, 101, S82–S88.

94. Valerio, A.; D'Antona, G.; Nisoli, E. Branched-chain amino acids, mitochondrial biogenesis, and healthspan: An evolutionary perspective. Aging 2011, 3, 464–478.

95. Newgard, C.B.; An, J.; Bain, J.R.; Muehlbauer, M.J.; Stevens, R.D.; Lien, L.F.; Haqq, A.M.; Shah, S.H.; Arlotto, M.; Slentz, C.A.; et al. A Branched-Chain Amino Acid-Related Metabolic Signature that Differentiates Obese and Lean Humans and Contributes to Insulin Resistance. Cell Metab. 2009, 9, 311–326.

96. Wang, T.J.; Larson, M.G.; Vasan, R.S.; Cheng, S.; Rhee, E.P.; McCabe, E.; Lewis, G.D.; Fox, C.S.; Jacques, P.F.; Fernandez, C.; et al. Metabolite profiles and the risk of developing diabetes. Nat. Med. 2011, 17, 448–453.

97. Patti, M.E.; Brambilla, E.; Luzi, L.; Landaker, E.J.; Kahn, C.R. Bidirectional modulation of insulin action by amino acids. J. Clin. Invest. 1998, 101, 1519–1529.

98. Tremblay, F.; Jacques, H.; Marette, A. Modulation of insulin action by dietary proteins and amino acids: Role of the mammalian target of rapamycin nutrient sensing pathway. Curr. Opin. Clin. Nutr. Metab. Care 2005, 8, 457–462.

99. Tremblay, F.; Marette, A. Amino acid and insulin signaling via the mTOR/p70 S6 kinase pathway. A negative feedback mechanism leading to insulin resistance in skeletal muscle cells. J. Biol. Chem. 2001, 276, 38052–38060.

Myofibrillar Protein Synthesis Following Ingestion of Soy Protein Isolate at Rest and After Resistance Exercise in Elderly Men

YIFAN YANG, TYLER A. CHURCHWARD-VENNE,
NICHOLAS A. BURD, LEIGH BREEN, MARK A. TARNOPOLSKY,
AND STUART M. PHILLIPS

6.1 INTRODUCTION

Ageing is associated with sarcopenia [1] that ultimately results from an imbalance between rates of muscle protein synthesis and breakdown. Both physical activity and nutrient availability represent potent anabolic stimuli for adult muscle, however, the ability of elderly muscle to mount a robust increase in myofibrillar protein synthesis (MPS) in response to amino acids [2,3] and resistance exercise [4] is attenuated compared to that seen in the young; a phenomenon termed 'anabolic resistance' [2]. Previous studies have shown that both protein dose [2,5,6] and source (i.e., plant

Myofibrillar Protein Synthesis Following Ingestion of Soy Protein Isolate at Rest and After Resistance Exercise in Elderly Men. © Yang Y, Churchward-Venne TA, Burd NA, Breen L, Tarnopolsky MA and Phillips SM. Nutrition & Metabolism **9**,57 (2012). Licensed under a Creative Commons Attribution 2.0 Generic License, http://creativecommons.org/licenses/by/2.0/.

vs. animal) [7-11] are important in determining the postprandial response of MPS, which may be of particular relevance to the elderly. For example, we have recently demonstrated greater increases in post-exercise MPS in the elderly following bolus ingestion of 40 g vs. 20 g of whey protein [6]; a finding in contrast to our data from young adults who show a maximal MPS response with 20 g protein and no further increase with 40 g [5]. Thus, it appears that higher doses of protein [6,12], and/or leucine [13,14] to promote a greater aminoacidemia or leucinemia [7] are required by the elderly to maximize the response of MPS to protein ingestion.

The mechanisms underpinning the differential capacity of proteins from different sources to support increased rates of protein synthesis are not fully understood [15]. Whey protein [7,9,10] and bovine milk [8] (~20% whey protein) appear to stimulate greater rates of muscle protein synthesis than do proteins such as micellar casein or soy both at rest and following resistance exercise. This is somewhat counter-intuitive given that soy, whey, and casein are all defined as high quality proteins based on their protein digestibility corrected amino acid scores (PDCAAS; for review see [16]). However, the digestion kinetics of these proteins is markedly different, and protein digestibility has been established as an important factor regulating whole-body protein synthesis and breakdown [17,18]. Both whey [18] and soy [19] are acid soluble, a characteristic that facilitates rapid digestion and results in a large but transient increase in aminoacidemia. These so-called 'fast' proteins induce a rapid aminoacidemia and appear to support greater increases in MPS. On the other hand 'slow' proteins, such as micellar casein (which clots in the acidic pH of the stomach) is slowly digested and induces a more moderate but sustained aminoacidemia than whey [7,10].

Knowledge of the capacity of proteins from different sources to stimulate MPS in the elderly is warranted in view of the importance of preserving skeletal muscle mass in ageing. Therefore, the aim of the current study was to examine the effects of different doses (20 g and 40 g) of soy protein isolate on MPS at rest and following the potent anabolic condition of resistance exercise in elderly men, and compare these findings to our previous work examining the effects of graded intakes of whey protein isolate on MPS in the elderly [6].

TABLE 1: Participant characteristics

Parameter	0 g (n=10)	W20 (n=10)	W40 (n=10)	S20 (n=10)	S40 (n=10)
Age (y)	71±5	72±5	70±4	72±6	70±5
Total body mass (kg)	78±13	81±9	81±12	78±11	77±9
Fat free mass (kg)	55±9	57±6	56±9	55±6	53±6
% Body fat	26±5	26±4	27±8	25±5	26±6
BMD (g·cm^2)	1.19±0.11	1.20±0.11	1.23±0.11	1.25±0.08	1.28±0.11
Height (m)	1.73±0.06	1.76±0.06	1.75±0.09	1.71±0.09	1.74±0.06
BMI (kg·m^2)	25.9±3.4	26.2±2.8	26.0±2.2	26.6±3.7	25.5±2.7
Systolic BP (mmHg)	136±15	134±19	129±14	124±13	127±12
Diastolic BP (mmHg)	80±10	72±8	78±5	73±9	72±8
Total SPBB score	11.7±0.5	11.3±0.7	11.6±0.7	11.8±0.4	11.4±1.0

Values are means±SD. BP: Blood pressure, SPBB: Short Physical Performance Battery. Total SPBB score calculated as the sum of walk test, chair stand and balance test.

6.2 METHODS

6.2.1 PARTICIPANTS

Thirty elderly men (age 71±5 y, BMI 26±3 kg·m^2) were recruited to participate in the study and were randomly assigned to one of three treatment groups that were counterbalanced for body mass, age, and self-reported physical activity levels. Participants were light-to-moderately active, non-smokers, non-diabetic, and considered generally healthy based on responses to a routine health screening questionnaire. Participants taking medications controlling blood pressure were allowed into the study. The characteristics of the whey protein treatment groups (W20, W40) have been reported previously [6], but are shown again in Table 1 along with the control (0 g) and soy protein treatment groups (S20, S40) for reader comparison. Participants were informed of the purpose of the study, the associated experimental procedures, and any potential risks prior to providing written consent. The study was approved by the local Health Sciences Research Ethics Board at McMaster University and conformed to standards for the use of human participants in research as outlined in the 5th Declaration of Helsinki and with current Canadian funding agency guidelines for use of human participants in research [20].

6.2.2 GENERAL DESIGN

The different groups of older men ingested 20 g or 40 g of soy protein isolate in beverage form after performing an acute bout of unilateral knee-extensor resistance exercise. Employing a unilateral exercise model allowed us to examine the effect of protein intake alone, and the interaction of exercise and protein intake within the same individual.

6.2.3 PRELIMINARY ASSESSMENTS

One week prior to the experimental infusion trial, body mass and composition were assessed via a dual energy X-ray absorptiometry (DXA) scan (Table 1). Physical performance was assessed using the Short Physical Performance Battery (SPPB) [21], consisting of a 3-4 m walk test, chair stand, and balance test (Total SPPB score presented in Table 1). Health parameters were also assessed and included systolic and diastolic blood pressure, resting heart rate, and the following blood parameters: fasting glucose, triglycerides, total cholesterol, high density lipoprotein (HDL), low density lipoprotein and ratio of total cholesterol to HDL. At least one week prior to the experimental infusion trial, participants underwent a strength test to determine their unilateral 10 repetition maximum (RM) on a standard knee extension machine as previously described [9].

6.2.4 DIETARY CONTROL

Participants were required to complete diet records prior to initiating the study to provide an estimate of habitual macronutrient intake as analyzed using a commercially available software program (Nutritionist V, First

Data Bank, San Bruno, CA). Reference lists for portion size estimates were provided to the study participants, who were instructed to record all food or drink consumed in a diet log during a 3-day period (i.e., 2 weekdays and 1 weekend day; see Additional file 1: Table S1). Two days prior to the trial, participants were supplied with pre-packaged diets that provided a moderate protein intake ($1.0\,g\cdot kg^{-1}\cdot d^{-1}$). Energy requirements for the controlled diets were estimated via the Harris-Benedict equation and were adjusted using an activity factor calculated for each individual subject based on their self reported physical activity. Body mass was monitored over the course of the controlled diet period to ensure participants were in energy balance. Participants were instructed to abstain from any strenuous exercise until after completion of the trial.

6.2.5 INFUSION PROTOCOL

Participants reported to the laboratory at ~0700 in a 10 h post-absorptive state. Upon arriving at the laboratory, a baseline breath sample was collected to measure $^{13}CO_2$ enrichment via isotope ratio mass spectrometry (BreathMat Plus; Finnigan MAT GmbH, Bremen, Germany). A plastic catheter was then inserted into an antecubital vein and a baseline blood sample was collected before initiating a 0.9% saline drip to keep the catheter patent for repeated blood sampling during the infusion trial. After baseline breath and blood samples were taken, a bout of unilateral knee-extensor resistance exercise was performed on a guided-motion knee extension machine. The exercise bout involved 3 sets, using a pre-determined load based on each participant's 10RM. Each set was completed within ~25 s with an interest rest interval of 2 min. Immediately following exercise, blood and breath samples were obtained and a second catheter was inserted into the contralateral antecubital vein to prime the bicarbonate pool with $NaH^{13}CO_2$ (2.35 μmol·kg). Thereafter, priming doses of [1-13 C] leucine (7.6 μmol·kg^{-1}) and L-ring-$^{13}C_6$ phenylalanine (2 μmol·kg^{-1}; 99 atom percent; Cambridge Isotopes, Andover, MA) were introduced, before a continuous infusion of L-[1-^{13}C] leucine (7.6 μmol·kg^{-1}·h^{-1}) and L-ring-$^{13}C_6$ phenylalanine was initiated (0.05 μmol·kg^{-1}·min^{-1}). Arterialized blood samples were obtained by wrapping the forearm in a heating blanket (45°C) for the duration of the infusion; a procedure we have found completely arterializes venous blood sampled from a hand vein. Blood samples were processed as previously described [5]. Immediately after post-exercise blood and breath samples had been obtained, participants consumed water (0 g) or a drink containing 20 g or 40 g of either whey or soy protein isolate (W20, W40, S20, S40) dissolved in 400 mL water. The whey protein was generously donated by PGP International (IWPI 9500, California, USA), while the soy protein was generously donated by the Solae Company (SUPRO 660-IP, St Louis, MO). The amino acid composition of both the whey and soy protein drinks is provided in Additional file 2: Table S2. On the basis of a leucine content of 10% in whey and 8% in soy, and a phenylalanine content of 3% in whey and 5% in soy protein, drinks were enriched to 8% with [1-^{13}C] leucine and 8% with ^{13}C$_6$ phenylalanine to minimize disturbances in isotopic steady state; an approach that we have validated [22]. Complete drink consumption was considered $t = 0$ min and the isotopic infusion was continued until $t = 240$ min. During the remainder of the infusion, arterialized blood and breath samples were obtained to confirm steady state and measure leucine oxidation and MPS as previously described [5,8]. At the end of the infusion ($t = 240$ min) muscle biopsies were obtained (described below).

6.2.6 MUSCLE BIOPSY SAMPLING

Muscle biopsy samples were obtained from the vastus lateralis muscle from both exercise and non-exercised legs using a 5-mm Bergstrm needle (modified for manual suction), under 2% local anaesthesia by xylocaine. Muscle biopsies were freed from any visible blood, fat, and connective tissue and rapidly frozen in liquid nitrogen until further analysis.

6.2.7 BLOOD ANALYSES

Plasma L-ring^{13}C$_6$ phenylalanine enrichments were determined as previously described [23]. Blood amino acid concentrations were analyzed by

HPLC as previously described [24]. Plasma insulin was measured using a commercially available immunoassay kit (ALPCO Diagnostics, Salem, NH, USA) following the manufacturer instructions.

6.2.8 MUSCLE ANALYSES

Myofibrillar enriched protein fractions were isolated from ~30 mg of wet muscle as described previously [25]. Intracellular amino acids (IC) were isolated from a separate piece of wet muscle (~25 mg) as previously described [26].

6.2.9 CALCULATIONS

The fractional synthetic rates (FSR) of myofibrillar proteins were calculated using the standard precursor-product method:

$$FSR\ (\%h^{-1}) = E_{p2} - E_{p1}/E_{ic} \times 1/t \times 100 \qquad (1)$$

where E_{p2} and E_{p1} are the protein bound enrichments from muscle biopsies at 240 min and baseline plasma proteins, respectively. The difference represents the change in bound protein enrichment between two time points; E_{ic} is the mean intracellular phenylalanine enrichment from the biopsies; and t is the tracer incorporation time. The utilization of 'tracer naive' subjects allowed us to use the pre-infusion blood sample (i.e., mixed plasma protein fraction) as a surrogate baseline enrichment of muscle protein; an approach we have previously validated [26] and that has been validated by others [27]. Previously, others have used a pre-infusion muscle biopsy and found equivalent rates of muscle protein synthesis and shown such an approach [28] to be valid. We have found baseline plasma enrichment to be equivalent to that of pre-infused muscle (unpublished results), indicating that there is little reason in using a pre-infusion biopsy over a blood sample for baseline enrichment.

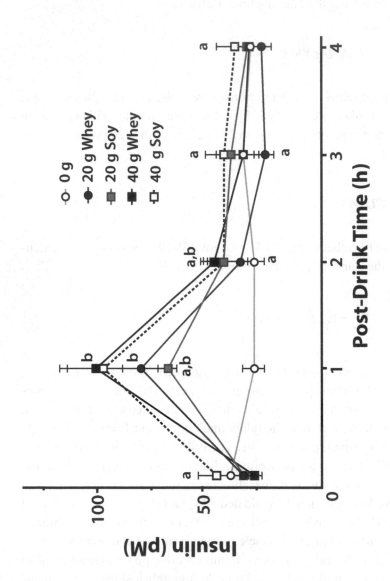

FIGURE 1: Plasma insulin in the 0 g protein group, and in the whey (W20, W40) and soy (S20, S40) groups. Means with different letters are significantly different from time 0 and from each other (P<0.05). Data are means±SD.

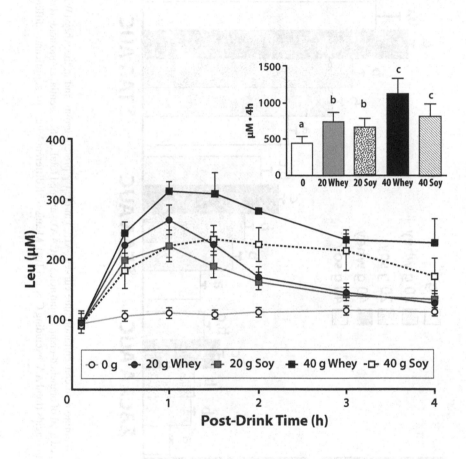

FIGURE 2: Blood leucine in the 0 g protein group, and in the whey (W20, W40) and soy (S20, S40) groups (no statistical analysis is shown except on the area under the concentration-time curve, see inset). Means with different letters are significantly different (P < 0.05). Data are means ± SD.

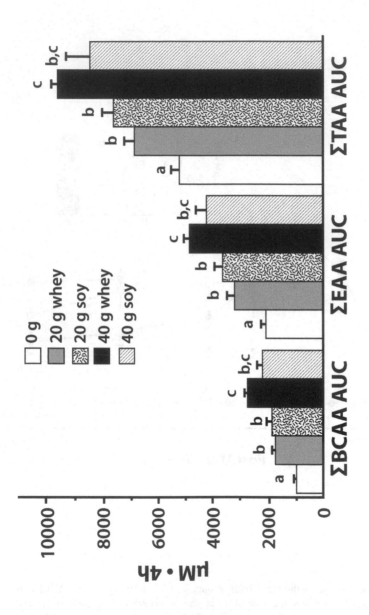

FIGURE 3: Area under the blood amino acid concentration time curves (AUC) in the 0 g protein group, and in the whey (W20, W40) and soy (S20, S40) groups for summed total of branched-chain amino acids (BCAA), summed total of the essential amino acids (EAA, including His) and summed total of all amino acids (total AA, excluding Cys and Trp). Means with different letters are significantly different (P<0.05). Data are means±SD.

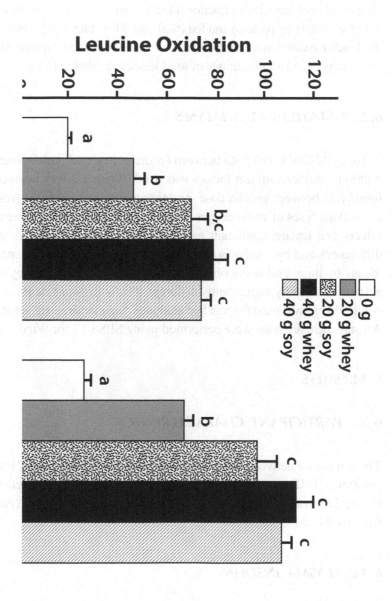

FIGURE 4: Whole-body leucine oxidation at different protein doses expressed relative to total body weight (left) and to lean body mass (right) in the 0g protein group, and in the whey (W20, W40) and soy (S20, S40) groups. Means with different letters are significantly different (P<0.05). Data are means±SD.

Leucine oxidation was calculated as described in our previous publications [5,8] from the appearance of the ^{13}C-label in expired CO_2 using the reciprocal pool model with fractional bicarbonate retention factors of 0.7 and 0.83 for fasted (0 g protein) and fed (S20 and S40) states [29]. The area under the leucine oxidation by time curve was calculated using GraphPad Prism 5 (San Diego, CA) as an estimate of total leucine oxidation [5,8].

6.2.10 STATISTICAL ANALYSES

A 3-way ANOVA with both between (protein dose and protein source), and within (condition) subject factors was used. When a 3-way interaction was found (i.e. between protein dose, protein source, and condition (rest vs. exercise)) analyses of variance was used to examine individual time and dose effects and isolate significant pairwise differences by calculating critical differences and by comparisons of means accounting for differences in the means by time. Following observation of a significant F ratio by ANOVA, a Tukey's honestly significantly different test, with adjustment for multiple comparisons, was used for post hoc analyses. Significance was set at $P \leq 0.05$. All statistical analyses were performed using SPSS 17 for Windows.

6.3 RESULTS

6.3.1 PARTICIPANT CHARACTERISTICS

There were no between-group differences in age, body weight, body composition, SPBB or other subject characteristics (Table 1). Dietary intake for the 2 day run-in prior to the study was similar for all groups (Additional file 1: Table S1).

6.3.2 PLASMA INSULIN

Plasma insulin concentration was similar for all groups at 0, 3 and 4 h post-drink. At 1 h post-drink, insulin concentration had increased by

~2.6- and 4-fold for W20 and W40, and ~2.2 fold for both S20 and S40 (Figure 1).

6.3.3 PLASMA AMINO ACIDS

Peak blood leucine concentration occurred between 1.0-1.5 h post-drink for S20, W20, and W40, but occurred at ~1.5-2.0 h for S40 (Figure 2). Higher peak amplitudes in blood leucinemia were achieved following whey as compared to soy protein regardless of dose (P<0.05). Area under the curve (AUC) for leucine increased in a stepwise manner from 20 g to 40 g of protein with no difference between protein sources (Figure 2 Inset). Blood BCAA, EAA, and Total amino acids increased with protein ingestion (i.e., versus 0 g), however there were no differences between protein sources (Figure 3). While the AUC for leucine, BCAA, EAA, and Total amino acids was greater for W40 vs. W20, there were no significant differences in these amino acid concentrations between S40 vs. S20.

6.3.4 WHOLE-BODY LEUCINE OXIDATION

Whole-body leucine oxidation AUC increased with protein intake. When expressed relative to lean body mass, the increase in whole body leucine oxidation for S20 was significantly greater than W20 (P=0.002). There were no differences in leucine oxidation between S40 and W40 (Figure 4).

6.3.5 MYOFIBRILLAR PROTEIN FRACTIONAL SYNTHETIC RATE (FSR)

Myofibrillar FSR in the non-exercise rested leg (fed only) was unchanged in response to ingestion of soy protein in both the S20 and S40 group, but increased in response to whey in both the W20 and W40 group (Figure 5). As such, MPS was significantly greater following whey vs. soy protein with ingestion of both 20 g and 40 g protein (both P<0.005). In the exercised condition, myofibrillar FSR was no different for S20, but increased for S40 when compared to the 0 g group. However, the response of MPS to soy was less than that of whey at both protein doses in the exercised condition (both P<0.001; Figure 5).

6.4 DISCUSSION

In the present study, we show that ingestion of 20 g (S20) and 40 g (S40) of soy protein isolate does not stimulate increased rates of MPS under resting conditions in the elderly. However, when combined with the potent anabolic stimulus of resistance exercise, 40 g but not 20 g, of soy protein isolate has a modest effect on increasing post-exercise rates of MPS when compared to a group who performed resistance exercise without subsequent protein intake (Figure 5). These data are in contrast to what was observed following equivalent doses of whey protein, as 20 g (W20) effectively stimulated MPS at rest, while both 20 g and 40 g (W40) increased post-exercise MPS in a stepwise manner (i.e. 40 g>20 g) [6]. Thus, when comparing our current findings on soy protein with our previous work examining graded doses of whey protein [6], soy appears less effective than whey protein at promoting increases in MPS in the elderly (Figure 5). Further, our results confirm that the elderly benefit from significantly greater doses of protein after exercise [6,12] than do the young, who we have shown mount a maximal MPS response with ingestion of ~20 g protein [5] or ~10 g EAA [2].

We have previously reported that soy protein is less effective than whey [9] and bovine milk protein [8] at increasing rates of post-exercise muscle protein synthesis in young subjects. Whey protein appears superior in its ability to stimulate muscle protein synthesis not only when compared to soy, but also when compared to other dairy protein sources such as intact [7,9,10] or hydrolyzed [10] casein. The mechanism(s) underpinning differences in the capacity of these proteins to support increased rates of MPS has not been fully elucidated. Previous research in rats reported greater increases in the phosphorylation status of mTOR(Ser 2448) and p70S6k (Thr 389), critical proteins involved in regulating translation initiation of protein synthesis, following whey compared with soy protein intake after endurance exercise [30]. Other important factors may relate to important differences in the leucine content of the respective proteins (~12% in whey and ~8% in soy) [16], and/or to differences in their digestion/absorption kinetics and the subsequent aminoacidemia [17,18,31]. For example, protein digestibility has been established as an important factor regulating whole-body protein synthesis and breakdown [17,18]; rapidly digested proteins have been shown to elicit a large increase in whole-body protein synthesis, whereas 'slow' proteins reduce rates of whole-body proteolysis [17,18,32]. More recent work has extended these findings at the whole-body level by showing that a fast protein, such as whey, also stimulates greater rates of skeletal muscle protein synthesis than does a slow protein, such as casein, both in both young and elderly subjects [7,9,10]. However, although whey and soy are relatively rapidly digested dietary proteins [19,33], previous studies have demonstrated that the amino acids from soy are partitioned for use within the body by more rapidly turning-over gut (i.e. splanchnic) proteins, and are converted to urea to a greater extent than amino acids from dairy based proteins which are partitioned to the periphery for use by skeletal muscle tissue [19,34].

In the present study, we observed protein source-dependent differences in rates of leucine oxidation (Figure 4). When expressed relative to lean body mass, rates of leucine oxidation were significantly greater for S20 than W20 (Figure 4). The higher rates of leucine oxidation in S20 vs. W20 suggest that a greater proportion of the amino acids from soy protein were diverted towards oxidation, and were thus unavailable as substrate for protein synthesis. Overall, although they are considered to be equivalent high quality proteins from the perspective of the truncated PDCAAS scoring system [16], there are clearly important differences in the capacity of soy and whey protein to stimulate MPS and promote anabolism. This point is of particular importance to the elderly in whom preserving skeletal muscle mass is of importance. Previous work showing that nitrogen balance is attainable with long-term diets containing moderate amounts of soy [35] would appear to be incongruent with our data; however, these data [35] are confounded by weight loss in a number of the subjects and due to the age of subjects in this study not being entirely comparable. Our data would, in contrast to previous conclusions regarding the adequacy of soy protein [35-37], suggest that long-term consumption of soy protein may not attenuate sarcopenic muscle loss.

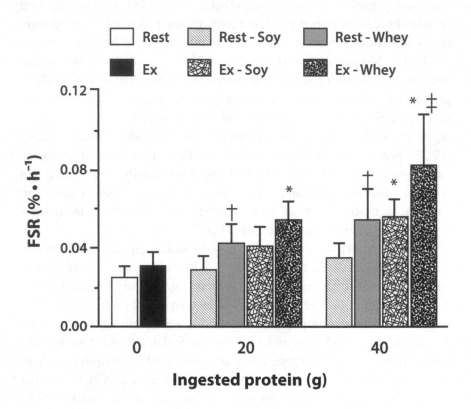

FIGURE 5: Myofibrillar protein fractional synthetic rate (%·h-1) for whey and soy (20 g and 40 g) groups and a group who consumed no protein (0 g) at rest and following resistance exercise (Ex) as described. There was a significant dose by condition by protein source interaction (P<0.001). † Indicates a significant (P<0.05) difference from the 0 g dose within the Rest condition; * indicates a significant (P<0.05) difference from the 0 g dose within the Ex condition; and ‡ indicates a significant (P<0.05) difference from the 20 g Ex condition within the same protein source. Data are means ± SD.

The mechanisms underpinning the 'anabolic resistance' of elderly muscle to nutrient provision are not entirely clear. Given the results of the current study, and previous studies demonstrating that MPS responds favorably to higher doses of protein in the elderly [6,12] as compared to the young [5], it appears that the muscle of older persons has a higher anabolic aminoacidemic 'threshold' [6,9,38] that can be surpassed by ingesting either greater quantities of protein/amino acids or possibly greater leucine [13]. The greater rates of MPS observed with equivalent doses of whey as compared to soy protein suggest that protein source is an important factor in reaching and surpassing the anabolic threshold (Figure 5). The branched chain amino acid leucine has been shown to be a key activator of muscle protein synthesis through its ability to regulate mRNA translation initiation through the mTOR signaling pathway [39,40]. For example, Katsanos and colleagues [13] reported that while 6.7 g of EAA containing ~26% leucine failed to stimulate MPS in the elderly, increasing the leucine content to ~41% increased MPS in the elderly such that measured rates were not different from that seen in the young. Based on results from the present study, there were no protein source dependent differences in leucine area under the curve (AUC) at either the 20 g or 40 g dose (Figure 3), however, the temporal response of blood leucine was different following whey and soy at both protein doses (Figure 2) with the response of whey being greater in amplitude than that observed following soy. To overcome the confounding influence of amino acid composition when comparing different proteins, we recently manipulated the pattern of postprandial aminoacidemia using a bolus versus a pulsed feeding pattern with whey protein [41]. Despite equivalent leucine and EAA AUC (i.e., net exposure) the bolus feeding pattern and the associated rapid aminoacidemia stimulated greater rates of post-exercise MPS than pulse feeding, which elicited a moderate but sustained rise in aminoacidemia [41]. Further, supplementation of soy protein with the BCAA has been shown to increase the anabolic effect of this protein in both the elderly and clinical COPD patients [42]. Thus, the higher leucine content and more rapid leucinemia with whey as opposed to soy may in part explain the observed differences in resting and post-exercise MPS between the two proteins.

In summary, we report that soy protein isolate is relatively ineffective in its capacity to stimulate MPS in the elderly when compared to whey protein. The mechanisms underpinning the reduced anabolic effect of soy as compared to whey likely relate to its relatively lower leucine content (~12% in whey and ~8% in soy) [16] and reduced leucinemia as a result of subtle differences in digestion/absorption between soy and whey protein. It is unlikely these differences have a marked impact on protein nutrition in all but the elderly or clinical populations [42]. Differences in postprandial amino acid oxidation rates may also be important as lower doses of soy (S20) resulted in greater increases in leucine oxidation than equivalent doses of whey protein. Our results have implications for nutrient formulations designed to support increased muscle protein anabolism in the elderly and suggest that whey protein offers clear advantages to soy protein in its capacity to support both rested and post-exercise increases in MPS.

REFERENCES

1.　Rosenberg IH: Sarcopenia: origins and clinical relevance. Clin Geriatr Med 2011, 27:337-339.
2.　Cuthbertson D, Smith K, Babraj J, Leese G, Waddell T, Atherton P, Wackerhage H, Taylor PM, Rennie MJ: Anabolic signaling deficits underlie amino acid resistance of wasting, aging muscle. The FASEB J: Off Publ of the Fed of Am Soc for Exp Biol 2005, 19:422-424.
3.　Volpi E, Mittendorfer B, Rasmussen BB, Wolfe RR: The response of muscle protein anabolism to combined hyperaminoacidemia and glucose-induced hyperinsulinemia is impaired in the elderly. J Clin Endocrinol Metabol 2000, 85:4481-4490.
4.　Kumar V, Selby A, Rankin D, Patel R, Atherton P, Hildebrandt W, Williams J, Smith K, Seynnes O, Hiscock N, Rennie MJ: Age-related differences in the dose-response relationship of muscle protein synthesis to resistance exercise in young and old men. J Physiol 2009, 587:211-217.
5.　Moore DR, Robinson MJ, Fry JL, Tang JE, Glover EI, Wilkinson SB, Prior T, Tarnopolsky MA, Phillips SM: Ingested protein dose response of muscle and albumin protein synthesis after resistance exercise in young men. Am J Clin Nutr 2009, 89:161-168.
6.　Yang Y, Breen L, Burd NA, Hector AJ, Churchward-Venne TA, Josse AR, Tarnopolsky MA, Phillips SM: Resistance exercise enhances myofibrillar protein synthesis with graded intakes of whey protein in older men. Br J Nutr 2012, 1-9.
7.　Burd NA, Yang Y, Moore DR, Tang JE, Tarnopolsky MA, Phillips SM: Greater stimulation of myofibrillar protein synthesis with ingestion of whey protein isolate v. micellar casein at rest and after resistance exercise in elderly men. Br J Nutr 2012, 1-5.
8.　Wilkinson SB, Tarnopolsky MA, Macdonald MJ, Macdonald JR, Armstrong D, Phillips SM: Consumption of fluid skim milk promotes greater muscle protein ac-

cretion after resistance exercise than does consumption of an isonitrogenous and isoenergetic soy-protein beverage. Am J Clin Nutr 2007, 85:1031-1040.

9. Tang JE, Moore DR, Kujbida GW, Tarnopolsky MA, Phillips SM: Ingestion of whey hydrolysate, casein, or soy protein isolate: effects on mixed muscle protein synthesis at rest and following resistance exercise in young men. J Appl Physiol 2009, 107:987-992.

10. Pennings B, Boirie Y, Senden JM, Gijsen AP, Kuipers H, van Loon LJ: Whey protein stimulates postprandial muscle protein accretion more effectively than do casein and casein hydrolysate in older men. Am J Clin Nutr 2011, 93:997-1005.

11. Reitelseder S, Agergaard J, Doessing S, Helmark IC, Lund P, Kristensen NB, Frystyk J, Flyvbjerg A, Schjerling P, van Hall G, et al.: Whey and casein labeled with L-[1-13 C]leucine and muscle protein synthesis: effect of resistance exercise and protein ingestion. Am J Physiol Endocrinol Metab 2011, 300:E231-242.

12. Pennings B, Groen BB, de Lange A, Gijsen AP, Zorenc AH, Senden JM, van Loon LJ: Amino acid absorption and subsequent muscle protein accretion following graded intakes of whey protein in elderly men. American journal of physiology Endocrinology and metabolism. 2012.

13. Katsanos CS, Kobayashi H, Sheffield-Moore M, Aarsland A, Wolfe RR: A high proportion of leucine is required for optimal stimulation of the rate of muscle protein synthesis by essential amino acids in the elderly. Am J Physiol Endocrinol Metab 2006, 291:E381-387.

14. Rieu I, Balage M, Sornet C, Giraudet C, Pujos E, Grizard J, Mosoni L, Dardevet D: Leucine supplementation improves muscle protein synthesis in elderly men independently of hyperaminoacidaemia. J Physiol 2006, 575:305-315.

15. Hayes A, Cribb PJ: Effect of whey protein isolate on strength, body composition and muscle hypertrophy during resistance training. Curr Opin Clin Nutr Metab Care 2008, 11:40-44.

16. Phillips SM, Tang JE, Moore DR: The role of milk- and soy-based protein in support of muscle protein synthesis and muscle protein accretion in young and elderly persons. J Am Coll Nutr 2009, 28:343-354.

17. Dangin M, Boirie Y, Garcia-Rodenas C, Gachon P, Fauquant J, Callier P, Ballevre O, Beaufrere B: The digestion rate of protein is an independent regulating factor of postprandial protein retention. Am J Physiol Endocrinol Metab 2001, 280:E340-348.

18. Boirie Y, Dangin M, Gachon P, Vasson MP, Maubois JL, Beaufrere B: Slow and fast dietary proteins differently modulate postprandial protein accretion. Proc Natl Acad Sci USA 1997, 94:14930-14935.

19. Bos C, Metges CC, Gaudichon C, Petzke KJ, Pueyo ME, Morens C, Everwand J, Benamouzig R, Tome D: Postprandial kinetics of dietary amino acids are the main determinant of their metabolism after soy or milk protein ingestion in humans. J Nutr 2003, 133:1308-1315.

20. Canadian Institutes of Health Research NSaERCoC, Social Sciences and Humanities Research Council of Canada, Tri-Council Policy Statement: Ethical Conduct for Research Involving Humans 2010.

21. Fisher S, Ottenbacher KJ, Goodwin JS, Graham J, Ostir GV: Short Physical Performance Battery in Hospitalized Older Adults. Aging Clin Exp Res 2009, 21:445.

22. Burd NA, West DW, Rerecich T, Prior T, Baker SK, Phillips SM: Validation of a single biopsy approach and bolus protein feeding to determine myofibrillar protein synthesis in stable isotope tracer studies. Nutr Metab 8

23. Glover EI, Phillips SM, Oates BR, Tang JE, Tarnopolsky MA, Selby A, Smith K, Rennie MJ: Immobilization induces anabolic resistance in human myofibrillar protein synthesis with low and high dose amino acid infusion. J Physiol 2008., 586

24. Moore DR, Phillips SM, Babraj JA, Smith K, Rennie MJ: Myofibrillar and collagen protein synthesis in human skeletal muscle in young men after maximal shortening and lengthening contractions. Am J Physiol Endocrinol Metab 2005, 288:E1153-1159.

25. Moore DR, Tang JE, Burd NA, Rerecich T, Tarnopolsky MA, Phillips SM: Differential stimulation of myofibrillar and sarcoplasmic protein synthesis with protein ingestion at rest and after resistance exercise. J Physiol 2009, 587:897-904.

26. Burd NA, Holwerda AM, Selby KC, West DW, Staples AW, Cain NE, Cashaback JG, Potvin JR, Baker SK, Phillips SM: Resistance exercise volume affects myofibrillar protein synthesis and anabolic signalling molecule phosphorylation in young men. J Physiol 2010, 588:3119-3130.

27. Burd NA, Groen BB, Beelen M, Senden JM, Gijsen AP, van Loon LJ: The reliability of using the single-biopsy approach to assess basal muscle protein synthesis rates in vivo in humans. Metab: Clin and Exp 2011.

28. Smith GI, Villareal DT, Lambert CP, Reeds DN, Selma Mohammed B, Mittendorfer B: Timing of the initial muscle biopsy does not affect the measured muscle protein fractional synthesis rate during basal, postabsorptive conditions. J Appl Physiol 2010, 108:363-368.

29. Tarnopolsky MA, Atkinson SA, MacDougall JD, Senor BB, Lemon PW, Schwarcz H: Whole body leucine metabolism during and after resistance exercise in fed humans. Med Sci Sports Exerc 1991, 23:326-333.

30. Anthony TG, McDaniel BJ, Knoll P, Bunpo P, Paul GL, McNurlan MA: Feeding meals containing soy or whey protein after exercise stimulates protein synthesis and translation initiation in the skeletal muscle of male rats. J Nutr 2007, 137:357-362.

31. Dangin M, Boirie Y, Guillet C, Beaufrere B: Influence of the protein digestion rate on protein turnover in young and elderly subjects. J Nutr 2002, 132:3228S-3233S.

32. Dangin M, Guillet C, Garcia-Rodenas C, Gachon P, Bouteloup-Demange C, Reiffers-Magnani K, Fauquant J, Ballevre O, Beaufrere B: The rate of protein digestion affects protein gain differently during aging in humans. J Physiol 2003, 549:635-644.

33. Luiking YC, Engelen MP, Soeters PB, Boirie Y, Deutz NE: Differential metabolic effects of casein and soy protein meals on skeletal muscle in healthy volunteers. Clin Nutr 2011, 30:65-72.

34. Fouillet H, Mariotti F, Gaudichon C, Bos C, Tome D: Peripheral and splanchnic metabolism of dietary nitrogen are differently affected by the protein source in humans as assessed by compartmental modeling. J Nutr 2002, 132:125-133.

35. Young VR, Wayler A, Garza C, Steinke FH, Murray E, Rand WM, Scrimshaw NS: A long-term metabolic balance study in young men to assess the nutritional quality of an isolated soy protein and beef proteins. Am J Clin Nutr 1984, 39:8-15.

36. Young VR: Soy protein in relation to human protein and amino acid nutrition. J Am Diet Assoc 1991, 91:828-835.
37. Young VR, Puig M, Queiroz E, Scrimshaw NS, Rand WM: Evaluation of the protein quality of an isolated soy protein in young men: relative nitrogen requirements and effect of methionine supplementation. Am J Clin Nutr 1984, 39:16-24.
38. Tang JE, Phillips SM: Maximizing muscle protein anabolism: the role of protein quality. Curr Opin Clin Nutr Metab Care 2009, 12:66-71.
39. Crozier SJ, Kimball SR, Emmert SW, Anthony JC, Jefferson LS: Oral leucine administration stimulates protein synthesis in rat skeletal muscle. J Nutr 2005, 135:376-382.
40. Anthony JC, Yoshizawa F, Anthony TG, Vary TC, Jefferson LS, Kimball SR: Leucine stimulates translation initiation in skeletal muscle of postabsorptive rats via a rapamycin-sensitive pathway. J Nutr 2000, 130:2413-2419.
41. West DW, Burd NA, Coffey VG, Baker SK, Burke LM, Hawley JA, Moore DR, Stellingwerff T, Phillips SM: Rapid aminoacidemia enhances myofibrillar protein synthesis and anabolic intramuscular signaling responses after resistance exercise. Am J Clin Nutr 2011, 94:795-803.
42. Engelen MP, Rutten EP, De Castro CL, Wouters EF, Schols AM, Deutz NE: Supplementation of soy protein with branched-chain amino acids alters protein metabolism in healthy elderly and even more in patients with chronic obstructive pulmonary disease. Am J Clin Nutr 2007, 85:431-439.

There are several supplemental files that are not available in this version of the article. To view this additional information, please use the citation on the first page of this chapter.

CHAPTER 7

The Link between Dietary Protein Intake, Skeletal Muscle Function, and Health in Older Adults

JAMIE I. BAUM AND ROBERT R. WOLFE

7.1 INTRODUCTION

The older adult population in the United States is a segment of unprecedented growth [1]. Longer life spans and aging of baby boomers will result in doubling of the population of Americans over the age of 65 over the next thirty years to reach almost 90 million people. By 2030, twenty percent of the US population will be comprised of older adults [1]. Chronic diseases such as heart disease, cancer and diabetes pose a great risk as people age. A majority of health care costs for older Americans can be attributed to treatment of chronic diseases [1]. People living with chronic disease(s) often experience diminished quality of life, which can be attributed to a long period of decline and disability [1].

Recycling of Pre-Washed Municipal Solid Waste Incinerator Fly Ash in the Manufacturing of Low Temperature Setting Geopolymer Materials. © *Baum JI and Wolfe RR.* Healthcare *3,3 (2015), doi:10.3390/healthcare3030529. Licensed under Creative Commons Attribution 4.0 International License, http://creativecommons.org/licenses/by/4.0/.*

Muscle mass, strength and function are progressively lost with aging [2]. A loss or reduction in skeletal muscle function often leads to increased morbidity and mortality either directly, or indirectly, via the development of secondary diseases such as diabetes, obesity and cardiovascular disease [2,3]. Sarcopenia is the age-associated loss of muscle mass and function. Sarcopenia can appear as early as age 40, increasing significantly over the age of 80, often resulting in 50% or greater loss of muscle strength [3]. The causes of sarcopenia appear to be multifactorial and include poor nutrition, diminished responsiveness to the normal anabolic effect from hormones and/or nutrients, and a sedentary lifestyle. The etiology of sarcopenia includes malnutrition, increased inflammatory cytokine production, oxidative stress, hormone reduction (e.g., growth hormone and testosterone), and decreased physical activity [4,5]. Maintaining skeletal muscle function throughout the lifespan into old age is essential for independent living and good health. The efficient activation of the mechanistic processes that regulate muscle development, growth, regeneration and metabolism is required for skeletal muscle to function at optimal levels [6].

Additional physical changes with aging may occur as a result of changes in skeletal muscle mass [2]. For example, body composition shifts as we age, resulting in a higher percentage of body fat and decreases in muscle mass with age [7]. This imbalance between muscle mass and fat mass occurs even in the absence of changes in body mass index (BMI) [7,8]. The prevalence of obesity among aging adults has also increased over the last several decades. For example, the prevalence of obesity among men aged 65–74 increased from 31.6% in 1999–2002 to 41.5% in 2007–2010. Between 2007 and 2010, approximately 35% of adults age 65 and over were obese [1].

Nutrition is an important modulator of health and function in older adults. Inadequate nutrition can contribute to the development of sarcopenia and obesity [2,5]. As life expectancy continues to rise, it is important to consider optimal nutritional recommendations that will improve health outcomes, quality of life and physical independence in older adults [1].

Several studies identify protein as a key nutrient for older adults (reviewed in [2,5]. Protein intake greater than the amount needed to avoid negative nitrogen balance may prevent sarcopenia [9], help maintain energy balance [10], improve bone health [11,12,13,14], cardiovascular function [15,16,17] and wound healing [18]. Benefits of increased protein intake may improve function and quality of life in healthy older adults, as well as improve the ability older patients to recover from disease and trauma [2].

7.2 DIETARY GUIDELINES FOR PROTEIN INTAKE: CURRENT RECOMMENDATIONS MAY UNDERESTIMATE OPTIMAL INTAKE IN OLDER ADULTS

7.2.1 CURRENT GUIDELINES FOR PROTEIN INTAKE

The most recent Dietary Guidelines for Americans were issued in 2010 by the USDA and were derived from the dietary reference intakes (DRIs) recommended by the Food and Nutrition Board of the Institute of Medicine [19,20]. The DRIs for macronutrients (Table 1) include an estimated average requirement (EAR), a recommended dietary allowance (RDA) and an acceptable macronutrient distribution range (AMDR) [19]. A comparison of the current dietary recommendations for dietary protein intake can be found in Table 1. In the case of daily protein intake, the EAR for dietary protein is 0.66 g/kg/day and the Food and Nutrition Board recommends an RDA of 0.8 g/kg/day for all adults >18 years of age, including older adults >65 years. The RDA for protein was based on all available studies that estimate the minimum protein intake necessary to avoid a progressive loss of lean body mass as determined by nitrogen balance [2,5]. The problem with relying only on results from nitrogen balance studies to define protein intake is that this method does not measure any physiological endpoints relevant to health, a point that was recognized by the Food and Nutrition Board [2]. In addition, the existing data were almost entirely from college-aged men [21]. Translating data from such a single age and gender group becomes an issue when trying to create universal recommendations for all adults over the age of 18, since older adults require greater nitrogen intake to maintain nitrogen balance compared to young adults [5]. Although the Food and Nutrition Board acknowledged this issue [20], they nonetheless relied on these data for the estimation of the RDA and did not differentiate the needs of young adults versus older adults [5].

The Food and Nutrition Board recognizes a distinction between the RDA and the level of protein intake needed for optimal health. The recommendation for the ADMRs (Table 1) includes a range of optimal protein intakes in the context of a complete diet (e.g., inclusive of additional macronutrient intake (e.g., carbohydrate and fat)), which makes the ADMR more relevant to normal dietary intake than the RDA [2]. The ADMR recommends that 10%–35% of daily energy intake come from protein [20].

Experts in the field of protein and aging recommend a protein intake between 1.2 and 2.0 g/kg/day or higher for older adults [2,5,22]. Clearly the RDA of 0.8 g/kg is well below these recommendations and reflects a value at the lowest end of the AMDR. Furthermore, current recommendations for protein intake for older adults fail to address needs of clinical conditions such as injury, hospitalization, surgery, etc., which are common in older adults and have been shown to require protein intake above recommended levels.

TABLE 1: Dietary protein intake recommendations.

Recommendation	Gram protein/kg body weight/day
EAR	0.66
RDA	0.8
AMDR *	~1.05–3.67

*EAR: Estimated Average Requirement; RDA: Recommended Dietary Allowance; AMDR: Acceptable Macronutrient Distribution Range; * Calculated assuming energy expenditure of 42 kcal/kg/day.*

7.2.2 QUANTITY VERSUS QUALITY OF PROTEIN INTAKE

It is not only the quantity of protein intake that is important for optimal health in older adults, it is the quality of the protein [23]. There are three important aspects to take into consideration when discussing protein quality: The characteristics of the specific protein, the food matrix in which the protein is consumed, and the characteristics of the individuals consuming the food (e.g., age, health status, physiological status and energy balance)

[23]. Current accepted methods for measuring protein quality do not consider the different metabolic roles the essential amino acids (EAA: leucine, valine, isoleucine, lysine, methionine, phenylalanine, tryptophan and histidine) play in the body beyond being the first limiting amino acid for growth or nitrogen balance [23]. Additionally, there are gaps in our knowledge regarding the influence of different protein sources on the metabolic health of older individuals.

Essential amino acids, especially the branched-chain amino acid leucine, are potent stimulators of muscle protein synthesis via the protein kinase mTORC1 (mammalian target of rapamycin complex 1) [24,25,26]. Several studies demonstrate that maximal stimulation of muscle protein synthesis is possible with 15 g of EAA (reviewed in [27]). This translates to ~35 g of high quality protein delivering ~15 g of EAA. A larger amount of lower quality protein, which contains a lower content of EAA, would be required to achieve the same functional benefits. The addition of non-essential amino acids to a supplement containing EAA does not result in additional stimulation of muscle protein synthesis [28], indicating that the quality of the protein, or its amino acid profile, is a key determinant of the functional potential of protein. This is supported by several studies demonstrating that ingestion of milk proteins stimulates muscle protein synthesis to a greater extent after resistance exercise compared to ingestion of soy protein [29,30,31,32]. The data from the Health, Aging and Body Composition study supports these findings [32]. These data show that intake of animal protein, not plant protein, was significantly associated with preservation of lean body mass over three years in older adults [32]. Individuals in the highest quintile of protein intake had 40% less loss in lean body mass than those in the lowest quintile of protein intake [32].

The majority of published results indicating a potential beneficial effect of increasing protein intake in older adults are either from epidemiological or short-term studies. Two recent publications by Tieland et al. [33,34] indicate that a significant proportion of the caloric intake above the RDA for caloric intake should be in the form of protein for older individuals. These conclusions are based on the results of randomized, double-blind, placebo-controlled, longer-term (24 weeks) studies. Improved physical performance in frail older adults results was reported when dietary protein intake was increased by twice-daily consumption of 15 g of milk protein.

These studies emphasize the importance of choosing the right protein for supplementation. Milk protein was used in these studies [33,34]. The proportion and profile of essential amino acids in milk protein make it a high-quality protein [35]. One important item to note with regard to these studies is that the control group was consuming 1.0 g protein/kg/day, which is 20% higher than the current RDA.

In addition, the difference in digestibility and bioavailability of a protein can impact the quantity of protein that needs to be ingested to meet metabolic needs. The speed of protein digestion and absorption of amino acids from the gut can influence whole body protein anabolism [36]. Proteins with differing amino acid profiles exhibit different digestion and absorption rates [36,37,38,39,40,41,42]. Amino acid availability depends directly on both the quality and quantity of the dietary source of nitrogen [37]. For example, the digestion and absorption rates of fast (e.g., whey) versus slow (e.g., casein) proteins need to be taken into consideration when developing protein recommendations. When young, healthy subjects were provided with either a whey protein meal (30 g) or a casein meal (43 g), both containing the same amount of leucine, and whole body protein anabolism was measured, the subjects consuming the whey (fast) protein meal had high, rapid increase in plasma amino acids, while subjects consuming the casein (slow) protein meal had a prolonged plateau of essential amino acids [40]. In addition, consumption of whey protein stimulated postprandial protein synthesis by 68%, while consumption of casein only stimulated protein synthesis by 31%. However, ingestion of casein inhibited whole body protein breakdown by 34%, while whey protein had no effect [40]. These results indicate that slow proteins, when adjusted for leucine content, may promote postprandial protein deposition by an inhibition of protein breakdown. When the influence of protein digestion rate on protein turnover was tested in older subjects, the opposite occurs [39]. When older men consumed 20 g of whey (2.5 g leucine), casein (1.7 g leucine) or casein hydrolysate (1.7 g leucine), the whey protein stimulated postprandial protein accretion more effectively than either casein or casein hydrolysate [39]. These results demonstrate that older and younger individuals respond differently to protein source and dose and highlight the need for separate protein recommendations for young and older adults.

Whether or not the amino acid source is an intact protein or a mixture of free amino acids can also influence the rate muscle protein synthesis [41]. For example, when older subjects were given either an EAA mixture (15 g) or a whey protein supplement (13.6 g) after an overnight fast, subjects consuming the EAA mixture had higher mixed muscle fractional synthetic rate [41]. The differing response could be due to differing leucine content between the supplements (EAA; 2.8 g leucine and Whey; 1.8 g leucine) or because the EAA supplement was composed of individual amino acids while the whey protein supplement was intact protein. This could influence rate of appearance of the amino acids into circulation and protein synthetic response. Another example is the form or texture of the protein itself, such as minced beef versus a beefsteak [42]. Older men consumed 135 g as either intrinsically labeled minced beef or beefsteak, which allowed direct assessment of the rate of appearance of dietary protein-derived amino acids from the gut into the circulation. Meat protein-derived phenylalanine appeared more rapidly in the circulation after minced beef than after beefsteak consumption. Whole-body protein balance was higher after consumption of the minced beef versus the beefsteak. Muscle protein synthesis did not differ between the two meals over a six-hour period [42]. These data indicate that format of protein during ingestion impacts digestion, absorption and rate of appearance of amino acids into circulation.

The interaction of protein with other substrates (e.g., fats, carbohydrates) may impact functionality of protein, especially at the molecular level. It is well established that the EAA, especially leucine, are essential for the functional benefits associated with increased protein intake in older adults, such as muscle protein synthesis [43,44,45,46]. However, most of these studies have tested either isolated EAA or protein independently of carbohydrates and fat. In the presence of the additional macronutrients, the pool of available EAA could be slower to appear when compared to studies examining EAA, alone, resulting in a decreased muscle protein synthesis response. Animal studies have shown that both fiber and fat can alter gastric emptying rate and delay the rate of appearance of EAA into the blood [47,48]. In healthy older adults, there has been no association with increased protein consumption and detrimental health effects. In contrast, detrimental effects have been associated with excess intake of both fat and carbohydrate. Therefore, not only should the beneficial effects of protein be considered when making dietary recommendations, but also the associated impact of lowering the ratio of fat and carbohydrate in the diet. Several studies have shown that a diet containing 30% protein or greater is beneficial in reducing cholesterol [49], improving glycemic regulation [43,44,50] and improving body composition (e.g., the ratio of lean to fat mass) [51]. These beneficial effects could be due to increasing dietary protein, alone, or a combination of increasing dietary protein and decreasing the amount of carbohydrate, however further investigation is needed.

Older adults are less responsive to low doses of amino acid intake compared to younger individuals [46]. However, this lack of responsiveness in older adults can be overcome with higher levels of amino acid consumption [46]. This is also reflected in studies comparing varying levels of protein consumption [52]. This suggests that the lack of muscle responsiveness to lower doses of amino acids in older adults can be overcome with a higher level of protein intake. The requirement for a larger dose of protein to generate responses in older adults similar to the responses in younger adults provides the support for a beneficial effect of increased protein in older populations [5].

7.2.3 TIMING OF PROTEIN INTAKE

Another issue related to defining the optimal protein recommendations for older adults is defining the optimal timing of protein intake throughout the day. Adults typically consume the majority of their protein (and energy) intake at dinner [53], 38 g versus 13 g of protein at breakfast. However, ingestion of more than 30 g of protein in a test meal does not further stimulate muscle protein synthesis [54]. Mamerow et al. [55] examined the effect of daily protein distribution on skeletal muscle protein synthesis in healthy young men and women for seven days. Ninety grams of protein was either distributed equally or unequally (60% of daily protein intake at dinner) throughout the day. They found even distribution of protein intake throughout day was more effective at stimulating 24 h protein synthesis compared to an uneven distribution [55]. This is further supported by Murphy et al. [56] who examined how dietary protein distribution affects muscle protein synthesis in overweight/obese older men under conditions of energy balance, energy restriction and energy restriction with the addition of resistance exercise. Protein was provided as a balanced distribution (25% of daily protein per meal, four times per day) or as a skewed pattern (7%:17%:72%:4% of daily protein per meal) for four weeks. They found that fed-state muscle fractional synthesis rates were higher following the balance protein intake pattern compared to the skewed protein intake pattern under energy balance and energy restricted conditions independent of resistance exercise [56].

A recent study published by Kim et al. [57] fails to confirm the importance of spreading protein intake out over the course of the day. Muscle protein synthesis was measured in healthy older adults after consumption of one of four differing daily patterns of protein intake: protein intake at the RDA level (0.8 g/kg/day), protein intake at two times the RDA level (1.5 g/kg/day), either evenly (33% protein at each meal) or unevenly (65% of protein at dinner) distributed throughout the day. In older adults, muscle protein synthesis was greater at two times the RDA than at RDA levels with no significant difference between meal patterns [57]. Likewise, Arnal et al. [58], conducted a study with older women fed protein at a dose of 1.7 g/kg fat free mass/day either spread over four meals throughout out the day or in one pulse feeding in which 80% of the daily protein intake was provided at the lunch meal for 14 days. Protein turnover and protein synthesis rates were higher in the pulse fed group compared to when protein was evenly distributed throughout the day [58]. However, in the study by Arnal et al. [58], the women only received a total of 65 g/d of protein with the distribution of protein spread throughout four meals with less than 20 g of protein per meal, while in the pulse treatment, the single large meal provided approximately 50 g of protein. This suggests that small meals spread throughout the day may fail to optimally stimulate muscle protein synthesis which could have negative effects on muscle mass. Additional studies compared pulse feeding (72% of daily protein at lunch) versus protein being evenly distributed over four daily meals in hospitalized older patients for six weeks [59,60]. These studies found that pulse feeding of protein increased postprandial amino acid bioavailability [59] and increased lean mass [60] compared to spreading protein intake throughout the day. These data may differ from the study conducted by Mamerow et al. [55] due to differences in study population (e.g., young versus older adults). In addition, the study by Mamerow et al. [55] only measured protein synthesis, which is only half of the equation of net protein gain.

7.3 FUNCTIONAL BENEFITS OF PROTEIN IN OLDER ADULTS

7.3.1 THE ROLE OF LEUCINE IN OLDER ADULTS

The current RDA for protein may not provide adequate EAA for optimal metabolic roles in adults. It is estimated that 7 to 12 grams of the branched-chain amino acid leucine are necessary per day [44] to see functional benefits beyond nitrogen balance, such as improved muscle function, glycemic regulation and mitochondrial function. Leucine may be more important in older adults than in young individuals. Availability of amino acids following a meal is lower in older adults when compared to young adults; therefore, it takes a greater protein content to elicit the same response postprandial. Several studies have demonstrated that older adults need a greater amount of leucine compared to young adults in order to achieve an increase in muscle protein synthesis [46,61,62].

7.3.2 INCREASED MUSCLE PROTEIN SYNTHESIS AND LEUCINE INTAKE IN OLDER ADULTS

The mechanism by which dietary protein affects muscle is through the stimulation of muscle protein synthesis by the absorbed amino acids consumed in the diet [27,63]. However, there appears to be an EAA threshold when it comes to stimulating muscle protein synthesis. Ingestion of relatively small amounts of essential amino acids (2.5, 5 or 10 g) appears to increase myofibrillar protein synthesis in a dose-dependent manner [64]. However, larger doses of EAA (20 to 40 g) fail to elicit an additional effect on protein synthesis in young and older subjects. Similar results were observed after ingestion of either 113 or 340 g of lean beef containing 10 or 30 g EAA, respectively [54]. Despite a three-fold increase in EAA content, there was no further increase in protein synthesis in either young or older subjects following consumption of 340 g versus 113 g of protein.

It is widely accepted that signaling via mTORC1 (mammalian target of rapamycin complex 1) is involved in the regulation of several anabolic processes including protein synthesis [26,65,66]. In skeletal muscle, amino acids signal through mTORC1 to regulate skeletal muscle protein synthesis [25,67,68,69]. The eukaryotic initiation factor 4E binding protein 1 (4E-BP1) and p70S6K, which play important roles in the initiation of mRNA translation, are downstream targets of mTORC1 [67,68,69]. Signals provided by EAA, especially leucine, are required for full activation of protein synthesis [25,67,70]. With increasing age, muscle may become resistant to the normal stimulatory effects postprandial leucine concentrations [46], which may result in reduced stimulation of the mTORC1 pathway. This could be due to a reduced sensitivity to leucine with age or to the fact that the older adults often have low dietary protein intake [5]. It is estimated that 38% of adult men and 41% of adult women have dietary protein intakes below the RDA. In addition, the decline in the rate of amino acid absorption followed by a decrease in insulin secretion may contribute to the insensitivity of muscle protein synthesis to amino acid-induced stimulation in older individuals [71,72].

Age-related muscle loss may involve a decreased response to EAA due to dysregulation of translation initiation factors. Older adults also have decreased levels of translational proteins related to muscle protein synthesis as compared to young adults [64,73]. Older adults have significantly less skeletal muscle mTORC1 and p70S6K protein levels compared to young adults [64]. In response to 10 g of EAA, mTORC1 phosphorylation, or activation, while significantly increased in skeletal muscle of older adults, is still significantly lower than younger adults. Guillet et al. [73] found that p70S6K phosphorylation is not stimulated in older adults after infusion with leucine. These findings are supported by Fry et al. [74] who found that older adults have significantly reduced phosphorylation of mTORC1 and translation initiation factors compared to young adults after a bout of resistance exercise. Gene expression of proteins associated with muscle protein synthesis also differs between young and older adults [74]. While no difference was found between young and old in the fasted state, there was a significant decrease in protein (REDD1, TSC1, TSC2 and IGF1 receptor) expression 3 and 6 h post exercise and leucine intervention in older adults versus young adults. In addition, after only seven days of bed rest, older adults had a reduced response to EAA ingestion resulting in no increase in muscle protein synthesis, activation of translation initiation factors (4E-BP1 and p70S6K) and no increase in amino acid transporters [75,76]. These findings are striking since pre-bed rest older adults had a significant increase in muscle protein synthesis following ingestion of EAA. These data are important because they demonstrate how quickly a hospital stay could impact skeletal muscle function.

It is also possible that the beneficial effect of protein intake on body composition is due to the stimulation of IGF-1 (insulin-like growth factor 1) secretion. Aging individuals have lower levels of IGF-1 [77], which could contribute to a decrease in protein synthesis rates and accentuate the loss of muscle mass leading to sarcopenia [78]. However, this may be able to be corrected by nutrition intervention since increased protein intake has been shown to increase circulating IGF-1 levels in older adults [79].

7.4 RISK FACTORS ASSOCIATED WITH LOW PROTEIN INTAKE

There are several risk factors associated with reduced protein intake in older adults [22]. These risk factors include reduced energy need and intake, reduced mobility, changes in food preference and food insecurity [22]. For example, as people age, their total daily energy needs decrease and their daily intake of dietary protein progressively declines, with approximately 8% of older women consuming protein below the estimated average requirement [53]. This could be due to underlying disease, changes in appetite and food preference or dental issues [80]. In addition, about 20% of homebound older adults have protein intakes less than the RDA [81], possibly attributed to difficulty acquiring and preparing food due to issues with mobility.

7.5 CLINICAL AND FINANCIAL IMPACT OF NUTRITION IN THE OLDER ADULTS

It is important to consider optimizing health care and those factors influencing health outcomes when determining dietary recommendations for dietary intake in older adults [2]. The cost of providing health care for one person over the age of 65 is three to five times more costly than for younger adults [1]. This could be because the average length of hospital stay for patients 65 and older is two days longer than younger age groups [1]. Ninety-five percent of health care costs for older Americans are for chronic diseases, which may be attributed, in part, to a loss of functional capacity related to reduced muscle mass [2]. Health care spending will increase by 25% by 2030, primarily due to an increasing older population [1].

Inadequate nutritional intake (e.g., low protein intake) is common in older adults and may explain the depleted muscle mass. Once in the hospital, physical inactivity combined with inadequate protein intake can result in additional loss of muscle mass which can delay recovery and contribute to higher readmission rates [5]. The breakdown of muscle in hospitalized older patients generates the amino acid precursors necessary for synthesis of proteins required for several body processes that are essential for recovery [82]. Many older patients do not have enough muscle mass to endogenously supply the amino acids they need, so the amino acids they need must be provided by dietary protein. However, protein needs are often not met in institutionalized and hospitalized settings [83,84]. There are several benefits associated with increased protein intake in the older individuals, including increased muscle protein synthesis [49,71,85,86] as well as aid in recovery from trauma [87,88] and surgery [79]. Taken together, higher protein intake improves the rate of recovery in older patients, which, in turn, may decrease the length of a hospital stay, indicating that dietary protein can play a central role in patient care [2].

7.6 CONCLUSIONS

There is sufficient evidence that protein intake higher than the current dietary recommendations (0.8 g/kg/day) is beneficial for most older adults. Higher protein intakes are associated with increased muscle protein synthesis, which is correlated with increased muscle mass and function. This, in turn, is linked to improved physical function. The evidence presented in this review supports the need for a higher RDA for protein for older adults in order for them to achieve optimal protein intake. Many older people suffering from chronic disease (e.g., diabetes, heart disease) have reduced appetite and do not consume adequate levels protein. Therefore, it is essential that they have adequate protein intake from high quality protein sources.

REFERENCES

1. Centers for Disease Control and Prevention, U.S. Department of Health and Human Services. The State of Aging and Health in America 2013; CreateSpace Independent Publishing Platform: Denver, CO, USA, 2013.
2. Wolfe, R.R. The role of dietary protein in optimizing muscle mass, function and health outcomes in older individuals. Br. J. Nutr. 2012, 108, S88–S93.
3. Arthur, S.T.; Cooley, I.D. The effect of physiological stimuli on sarcopenia; impact of notch and Wnt signaling on impaired aged skeletal muscle repair. Int. J. Biol. Sci. 2012, 8, 731–760.
4. Kim, T.N.; Choi, K.M. Sarcopenia: Definition, epidemiology, and pathophysiology. J. Bone Metab. 2013, 20, 1–10.
5. Wolfe, R.R.; Miller, S.L.; Miller, K.B. Optimal protein intake in the elderly. Clin. Nutr. 2008, 27, 675–684.
6. Guller, I.; Russell, A.P. MicroRNAs in skeletal muscle: Their role and regulation in development, disease and function. J. Physiol. 2010, 588, 4075–4087.
7. Chumlea, W.C.; Baumgartner, R.N.; Vellas, B.P. Anthropometry and body composition in the perspective of nutritional status in the elderly. Nutrition 1991, 7, 57–60.
8. Kim, T.N.; Won, J.C.; Kim, Y.J.; Lee, E.J.; Kim, M.K.; Park, M.S.; Lee, S.K.; Kim, J.M.; Ko, K.S.; Rhee, B.D. Serum adipocyte fatty acid-binding protein levels are independently associated with sarcopenic obesity. Diabetes Res. Clin. Pract. 2013, 101, 210–217.
9. Morais, J.A.; Chevalier, S.; Gougeon, R. Protein turnover and requirements in the healthy and frail elderly. J. Nutr. Health Aging 2006, 10, 272–283.
10. Wilson, M.M.; Purushothaman, R.; Morley, J.E. Effect of liquid dietary supplements on energy intake in the elderly. Am. J. Clin. Nutr. 2002, 75, 944–947.
11. Dawson-Hughes, B. Calcium and protein in bone health. Proc. Nutr. Soc. 2003, 62, 505–509.
12. Dawson-Hughes, B. Interaction of dietary calcium and protein in bone health in humans. J. Nutr. 2003, 133, 852S–854S.
13. Thorpe, M.P.; Jacobson, E.H.; Layman, D.K.; He, X.; Kris-Etherton, P.M.; Evans, E.M. A diet high in protein, dairy, and calcium attenuates bone loss over twelve months of weight loss and maintenance relative to a conventional high-carbohydrate diet in adults. J. Nutr. 2008, 138, 1096–1100.
14. Heaney, R.P.; Layman, D.K. Amount and type of protein influences bone health. Am. J. Clin. Nutr. 2008, 87, 1567S–1570S.

15. Hu, F.B.; Stampfer, M.J.; Manson, J.E.; Rimm, E.; Colditz, G.A.; Speizer, F.E.; Hennekens, C.H.; Willett, W.C. Dietary protein and risk of ischemic heart disease in women. Am. J. Clin. Nutr. 1999, 70, 221–227.

16. Obarzanek, E.; Velletri, P.A.; Cutler, J.A. Dietary protein and blood pressure. Jama 1996, 275, 1598–1603.

17. Stamler, J.; Elliott, P.; Kesteloot, H.; Nichols, R.; Claeys, G.; Dyer, A.R.; Stamler, R. Inverse relation of dietary protein markers with blood pressure. Findings for 10,020 men and women in the INTERSALT Study. INTERSALT Cooperative Research Group. INTERnational study of SALT and blood pressure. Circulation 1996, 94, 1629–1634.

18. Stratton, R.J.; Ek, A.C.; Engfer, M.; Moore, Z.; Rigby, P.; Wolfe, R.; Elia, M. Enteral nutritional support in prevention and treatment of pressure ulcers: A systematic review and meta-analysis. Ageing Res. Rev. 2005, 4, 422–450.

19. U.S. Department of Agriculture; U.S. Department of Health and Human Services. Dietary Guidelines for Americans, 7th ed.; U.S. Government Printing Office: Washington, DC, USA, 2010.

20. Trumbo, P.; Schlicker, S.; Yates, A.A.; Poos, M.; Food and Nutrition Board of the Institute of Medicine; The National Academies. Dietary Reference Intakes for Energy, Carbohydrate, Fiber, Fat, Fatty Acids, Cholesterol, Protein and Amino Acids; National Academy Press: Washington, DC, USA, 2002.

21. Campbell, W.W.; Crim, M.C.; Dallal, G.E.; Young, V.R.; Evans, W.J. Increased protein requirements in elderly people: New data and retrospective reassessments. Am. J. Clin. Nutr. 1994, 60, 501–509.

22. Volpi, E.; Campbell, W.W.; Dwyer, J.T.; Johnson, M.A.; Jensen, G.L.; Morley, J.E.; Wolfe, R.R. Is the optimal level of protein intake for older adults greater than the recommended dietary allowance? J. Gerontol. A Biol. Sci. Med. Sci. 2013, 68, 677–681.

23. Millward, D.J.; Layman, D.K.; Tome, D.; Schaafsma, G. Protein quality assessment: Impact of expanding understanding of protein and amino acid needs for optimal health. Am. J. Clin. Nutr. 2008, 87, 1576S–1581S.

24. Anthony, J.C.; Anthony, T.G.; Kimball, S.R.; Vary, T.C.; Jefferson, L.S. Orally administered leucine stimulates protein synthesis in skeletal muscle of postabsorptive rats in association with increased eIF4F formation. J. Nutr. 2000, 130, 139–145.

25. Anthony, J.C.; Yoshizawa, F.; Anthony, T.G.; Vary, T.C.; Jefferson, L.S.; Kimball, S.R. Leucine stimulates translation initiation in skeletal muscle of postabsorptive rats via a rapamycin-sensitive pathway. J. Nutr. 2000, 130, 2413–2419.

26. Gordon, B.S.; Kelleher, A.R.; Kimball, S.R. Regulation of muscle protein synthesis and the effects of catabolic states. Int. J. Biochem. Cell Biol. 2013, 45, 2147–2157.

27. Wolfe, R.R. Regulation of muscle protein by amino acids. J. Nutr. 2002, 132, 3219S–3224S.

28. Borsheim, E.; Tipton, K.D.; Wolf, S.E.; Wolfe, R.R. Essential amino acids and muscle protein recovery from resistance exercise. Am. J. Physiol. Endocrinol. Metab. 2002, 283, E648–E657.

29. Mitchell, C.J.; Della Gatta, P.A.; Petersen, A.C.; Cameron-Smith, D.; Markworth, J.F. Soy protein ingestion results in less prolonged p70S6 kinase phosphorylation

compared to whey protein after resistance exercise in older men. J. Int. Soc. Sports Nutr. 2015.

30. Phillips, S.M.; Tang, J.E.; Moore, D.R. The role of milk- and soy-based protein in support of muscle protein synthesis and muscle protein accretion in young and elderly persons. J. Am. Coll. Nutr. 2009, 28, 343–354.

31. Tang, J.E.; Moore, D.R.; Kujbida, G.W.; Tarnopolsky, M.A.; Phillips, S.M. Ingestion of whey hydrolysate, casein, or soy protein isolate: Effects on mixed muscle protein synthesis at rest and following resistance exercise in young men. J. Appl. Physiol. 1985 2009, 107, 987–992.

32. Houston, D.K.; Nicklas, B.J.; Ding, J.; Harris, T.B.; Tylavsky, F.A.; Newman, A.B.; Lee, J.S.; Sahyoun, N.R.; Visser, M.; Kritchevsky, S.B. Dietary protein intake is associated with lean mass change in older, community-dwelling adults: The Health, Aging, and Body Composition (Health ABC) Study. Am. J. Clin. Nutr. 2008, 87, 150–155.

33. Tieland, M.; van de Rest, O.; Dirks, M.L.; van der Zwaluw, N.; Mensink, M.; van Loon, L.J.; de Groot, L.C. Protein supplementation improves physical performance in frail elderly people: A randomized, double-blind, placebo-controlled trial. J. Am. Med. Dir. Assoc. 2012, 13, 720–726.

34. Tieland, M.; Dirks, M.L.; van der Zwaluw, N.; Verdijk, L.B.; van de Rest, O.; de Groot, L.C.; van Loon, L.J. Protein supplementation increases muscle mass gain during prolonged resistance-type exercise training in frail elderly people: A randomized, double-blind, placebo-controlled trial. J. Am. Med. Dir. Assoc. 2012, 13, 713–719.

35. Wolfe, R.R. Perspective: Optimal protein intake in the elderly. J. Am. Med. Dir. Assoc. 2013, 14, 65–66.

36. Boirie, Y.; Dangin, M.; Gachon, P.; Vasson, M.P.; Maubois, J.L.; Beaufrere, B. Slow and fast dietary proteins differently modulate postprandial protein accretion. Proc. Natl. Acad. Sci. USA 1997, 94, 14930–14935.

37. Dangin, M.; Boirie, Y.; Garcia-Rodenas, C.; Gachon, P.; Fauquant, J.; Callier, P.; Ballevre, O.; Beaufrere, B. The digestion rate of protein is an independent regulating factor of postprandial protein retention. Am. J. Physiol. Endocrinol. Metab. 2001, 280, E340–E348.

38. Dangin, M.; Boirie, Y.; Guillet, C.; Beaufrere, B. Influence of the protein digestion rate on protein turnover in young and elderly subjects. J. Nutr. 2002, 132, 3228S–3233S.

39. Pennings, B.; Boirie, Y.; Senden, J.M.; Gijsen, A.P.; Kuipers, H.; van Loon, L.J. Whey protein stimulates postprandial muscle protein accretion more effectively than do casein and casein hydrolysate in older men. Am. J. Clin. Nutr. 2011, 93, 997–1005.

40. Boirie, Y.; Gachon, P.; Beaufrere, B. Splanchnic and whole-body leucine kinetics in young and elderly men. Am. J. Clin. Nutr. 1997, 65, 489–495.

41. Paddon-Jones, D.; Sheffield-Moore, M.; Katsanos, C.S.; Zhang, X.J.; Wolfe, R.R. Differential stimulation of muscle protein synthesis in elderly humans following isocaloric ingestion of amino acids or whey protein. Exp. Gerontol. 2006, 41, 215–219.

42. Pennings, B.; Groen, B.B.; van Dijk, J.W.; de Lange, A.; Kiskini, A.; Kuklinski, M.; Senden, J.M.; van Loon, L.J. Minced beef is more rapidly digested and absorbed than beef steak, resulting in greater postprandial protein retention in older men. Am. J. Clin. Nutr. 2013, 98, 121–128.

43. Layman, D.K.; Baum, J.I. Dietary protein impact on glycemic control during weight loss. J. Nutr. 2004, 134, 968S–973S.

44. Layman, D.K. The role of leucine in weight loss diets and glucose homeostasis. J. Nutr. 2003, 133, 261S–267S.

45. Layman, D.K. Role of leucine in protein metabolism during exercise and recovery. Can. J. Appl. Physiol. 2002, 27, 646–663.

46. Katsanos, C.S.; Kobayashi, H.; Sheffield-Moore, M.; Aarsland, A.; Wolfe, R.R. A high proportion of leucine is required for optimal stimulation of the rate of muscle protein synthesis by essential amino acids in the elderly. Am. J. Physiol. Endocrinol. Metab. 2006, 291, E381–E387.

47. Anthony, J.C.; Reiter, A.K.; Anthony, T.G.; Crozier, S.J.; Lang, C.H.; MacLean, D.A.; Kimball, S.R.; Jefferson, L.S. Orally administered leucine enhances protein synthesis in skeletal muscle of diabetic rats in the absence of increases in 4E-BP1 or S6K1 phosphorylation. Diabetes 2002, 51, 928–936.

48. Norton, L.E.; Layman, D.K.; Bunpo, P.; Anthony, T.G.; Brana, D.V.; Garlick, P.J. The leucine content of a complete meal directs peak activation but not duration of skeletal muscle protein synthesis and mammalian target of rapamycin signaling in rats. J. Nutr. 2009, 139, 1103–1109.

49. Layman, D.K.; Boileau, R.A.; Erickson, D.J.; Painter, J.E.; Shiue, H.; Sather, C.; Christou, D.D. A reduced ratio of dietary carbohydrate to protein improves body composition and blood lipid profiles during weight loss in adult women. J. Nutr. 2003, 133, 411–417.

50. Layman, D.K.; Shiue, H.; Sather, C.; Erickson, D.J.; Baum, J. Increased dietary protein modifies glucose and insulin homeostasis in adult women during weight loss. J. Nutr. 2003, 133, 405–410.

51. Layman, D.K.; Evans, E.M.; Erickson, D.; Seyler, J.; Weber, J.; Bagshaw, D.; Griel, A.; Psota, T.; Kris-Etherton, P. A moderate-protein diet produces sustained weight loss and long-term changes in body composition and blood lipids in obese adults. J. Nutr. 2009, 139, 514–521.

52. Moore, D.R.; Churchward-Venne, T.A.; Witard, O.; Breen, L.; Burd, N.A.; Tipton, K.D.; Phillips, S.M. Protein ingestion to stimulate myofibrillar protein synthesis requires greater relative protein intakes in healthy older versus younger men. J. Gerontol. A Biol. Sci. Med. Sci. 2015, 70, 57–62.

53. Fulgoni, V.L., 3rd. Current protein intake in America: Analysis of the National Health and Nutrition Examination Survey, 2003–2004. Am. J. Clin. Nutr. 2008, 87, 1554S–1557S.

54. Symons, T.B.; Sheffield-Moore, M.; Wolfe, R.R.; Paddon-Jones, D. A moderate serving of high-quality protein maximally stimulates skeletal muscle protein synthesis in young and elderly subjects. J. Am. Diet. Assoc. 2009, 109, 1582–1586.

55. Mamerow, M.M.; Mettler, J.A.; English, K.L.; Casperson, S.L.; Arentson-Lantz, E.; Sheffield-Moore, M.; Layman, D.K.; Paddon-Jones, D. Dietary protein distribution

positively influences 24-h muscle protein synthesis in healthy adults. J. Nutr. 2014, 144, 876–880.

56. Murphy, C.H.; Churchward-Venne, T.A.; Mitchell, C.J.; Kolar, N.M.; Kassis, A.; Karagounis, L.G.; Burke, L.M.; Hawley, J.A.; Phillips, S.M. Hypoenergetic diet-induced reductions in myofibrillar protein synthesis are restored with resistance training and balanced daily protein ingestion in older men. Am. J. Physiol. Endocrinol. Metab. 2015, 308, E734–E743.

57. Kim, I.Y.; Schutzler, S.; Schrader, A.; Spencer, H.; Kortebein, P.; Deutz, N.E.; Wolfe, R.R.; Ferrando, A.A. Quantity of dietary protein intake, but not pattern of intake, affects net protein balance primarily through differences in protein synthesis in older adults. Am. J. Physiol. Endocrinol. Metab. 2015, 308, E21–E28.

58. Arnal, M.A.; Mosoni, L.; Boirie, Y.; Houlier, M.L.; Morin, L.; Verdier, E.; Ritz, P.; Antoine, J.M.; Prugnaud, J.; Beaufrere, B.; et al. Protein pulse feeding improves protein retention in elderly women. Am. J. Clin. Nutr. 1999, 69, 1202–1208.

59. Bouillanne, O.; Neveux, N.; Nicolis, I.; Curis, E.; Cynober, L.; Aussel, C. Long-lasting improved amino acid bioavailability associated with protein pulse feeding in hospitalized elderly patients: A randomized controlled trial. Nutrition 2014, 30, 544–550.

60. Bouillanne, O.; Curis, E.; Hamon-Vilcot, B.; Nicolis, I.; Chretien, P.; Schauer, N.; Vincent, J.P.; Cynober, L.; Aussel, C. Impact of protein pulse feeding on lean mass in malnourished and at-risk hospitalized elderly patients: A randomized controlled trial. Clin. Nutr. 2013, 32, 186–192.

61. Wall, B.T.; Hamer, H.M.; de Lange, A.; Kiskini, A.; Groen, B.B.; Senden, J.M.; Gijsen, A.P.; Verdijk, L.B.; van Loon, L.J. Leucine co-ingestion improves post-prandial muscle protein accretion in elderly men. Clin. Nutr. 2013, 32, 412–419.

62. Casperson, S.L.; Sheffield-Moore, M.; Hewlings, S.J.; Paddon-Jones, D. Leucine supplementation chronically improves muscle protein synthesis in older adults consuming the RDA for protein. Clin. Nutr. 2012, 31, 512–519.

63. Rasmussen, B.B.; Wolfe, R.R.; Volpi, E. Oral and intravenously administered amino acids produce similar effects on muscle protein synthesis in the elderly. J. Nutr. Health Aging 2002, 6, 358–362.

64. Cuthbertson, D.; Smith, K.; Babraj, J.; Leese, G.; Waddell, T.; Atherton, P.; Wackerhage, H.; Taylor, P.M.; Rennie, M.J. Anabolic signaling deficits underlie amino acid resistance of wasting, aging muscle. FASEB J. 2005, 19, 422–424.

65. Sakuma, K.; Aoi, W.; Yamaguchi, A. Current understanding of sarcopenia: Possible candidates modulating muscle mass. Pflugers Arch. 2015, 467, 213–229.

66. Sakuma, K.; Aoi, W.; Yamaguchi, A. The intriguing regulators of muscle mass in sarcopenia and muscular dystrophy. Front. Aging Neurosci. 2014.

67. Kimball, S.R.; Jefferson, L.S. Signaling pathways and molecular mechanisms through which branched-chain amino acids mediate translational control of protein synthesis. J. Nutr. 2006, 136, 227S–231S.

68. Kimball, S.R.; Jefferson, L.S. Role of amino acids in the translational control of protein synthesis in mammals. Semin. Cell Dev. Biol. 2005, 16, 21–27.

69. Kimball, S.R.; Jefferson, L.S. Regulation of global and specific mRNA translation by oral administration of branched-chain amino acids. Biochem. Biophys. Res. Commun. 2004, 313, 423–427.

70. Wilkinson, D.J.; Hossain, T.; Hill, D.S.; Phillips, B.E.; Crossland, H.; Williams, J.; Loughna, P.; Churchward-Venne, T.A.; Breen, L.; Phillips, S.M.; et al. Effects of leucine and its metabolite β-hydroxy-β-methylbutyrate on human skeletal muscle protein metabolism. J. Physiol. 2013, 591, 2911–2923.

71. Katsanos, C.S.; Kobayashi, H.; Sheffield-Moore, M.; Aarsland, A.; Wolfe, R.R. Aging is associated with diminished accretion of muscle proteins after the ingestion of a small bolus of essential amino acids. Am. J. Clin. Nutr. 2005, 82, 1065–1073.

72. Paddon-Jones, D.; Sheffield-Moore, M.; Creson, D.L.; Sanford, A.P.; Wolf, S.E.; Wolfe, R.R.; Ferrando, A.A. Hypercortisolemia alters muscle protein anabolism following ingestion of essential amino acids. Am. J. Physiol. Endocrinol. Metab. 2003, 284, E946–E953.

73. Guillet, C.; Prod'homme, M.; Balage, M.; Gachon, P.; Giraudet, C.; Morin, L.; Grizard, J.; Boirie, Y. Impaired anabolic response of muscle protein synthesis is associated with S6K1 dysregulation in elderly humans. FASEB J. 2004, 18, 1586–1587.

74. Fry, C.S.; Drummond, M.J.; Glynn, E.L.; Dickinson, J.M.; Gundermann, D.M.; Timmerman, K.L.; Walker, D.K.; Volpi, E.; Rasmussen, B.B. Skeletal muscle autophagy and protein breakdown following resistance exercise are similar in younger and older adults. J. Gerontol. A Biol. Sci. Med. Sci. 2013, 68, 599–607.

75. Drummond, M.J.; Miyazaki, M.; Dreyer, H.C.; Pennings, B.; Dhanani, S.; Volpi, E.; Esser, K.A.; Rasmussen, B.B. Expression of growth-related genes in young and older human skeletal muscle following an acute stimulation of protein synthesis. J. Appl. Physiol. 1985 2009, 106, 1403–1411.

76. Drummond, M.J.; Dickinson, J.M.; Fry, C.S.; Walker, D.K.; Gundermann, D.M.; Reidy, P.T.; Timmerman, K.L.; Markofski, M.M.; Paddon-Jones, D.; Rasmussen, B.B.; et al. Bed rest impairs skeletal muscle amino acid transporter expression, mTORC1 signaling, and protein synthesis in response to essential amino acids in older adults. Am. J. Physiol. Endocrinol. Metab. 2012, 302, E1113–E1122.

77. Kelijman, M. Age-related alterations of the growth hormone/insulin-like-growth-factor I axis. J. Am. Geriatr. Soc. 1991, 39, 295–307.

78. Ceda, G.P.; Dall'Aglio, E.; Maggio, M.; Lauretani, F.; Bandinelli, S.; Falzoi, C.; Grimaldi, W.; Ceresini, G.; Corradi, F.; Ferrucci, L.; et al. Clinical implications of the reduced activity of the GH-IGF-I axis in older men. J. Endocrinol. Investig. 2005, 28, 96–100.

79. Schurch, M.A.; Rizzoli, R.; Slosman, D.; Vadas, L.; Vergnaud, P.; Bonjour, J.P. Protein supplements increase serum insulin-like growth factor-I levels and attenuate proximal femur bone loss in patients with recent hip fracture. A randomized, double-blind, placebo-controlled trial. Ann. Intern. Med. 1998, 128, 801–809.

80. Morley, J.E. Anorexia, weight loss, and frailty. J. Am. Med. Dir. Assoc. 2010, 11, 225–228.

81. Locher, J.L.; Ritchie, C.S.; Roth, D.L.; Sen, B.; Vickers, K.S.; Vailas, L.I. Food choice among homebound older adults: Motivations and perceived barriers. J. Nutr. Health Aging 2009, 13, 659–664.

82. Wolfe, R.R. The underappreciated role of muscle in health and disease. Am. J. Clin. Nutr. 2006, 84, 475–482.

83. Keller, H.H. Malnutrition in institutionalized elderly: How and why? J. Am. Geriatr. Soc. 1993, 41, 1212–1218.

84. Morley, J.E. Anorexia of aging: Physiologic and pathologic. Am. J. Clin. Nutr. 1997, 66, 760–773.
85. Paddon-Jones, D.; Sheffield-Moore, M.; Zhang, X.J.; Volpi, E.; Wolf, S.E.; Aarsland, A.; Ferrando, A.A.; Wolfe, R.R. Amino acid ingestion improves muscle protein synthesis in the young and elderly. Am. J. Physiol. Endocrinol. Metab. 2004, 286, E321–E328.
86. Volpi, E.; Ferrando, A.A.; Yeckel, C.W.; Tipton, K.D.; Wolfe, R.R. Exogenous amino acids stimulate net muscle protein synthesis in the elderly. J. Clin. Investig. 1998, 101, 2000–2007.
87. Demling, R.H.; DeSanti, L. Increased protein intake during the recovery phase after severe burns increases body weight gain and muscle function. J. Burn Care Rehabil. 1998, 19, 161–168.
88. Hughes, M.S.; Kazmier, P.; Burd, T.A.; Anglen, J.; Stoker, A.M.; Kuroki, K.; Carson, W.L.; Cook, J.L. Enhanced fracture and soft-tissue healing by means of anabolic dietary supplementation. J. Bone Jt. Surg. Am. 2006, 88, 2386–2394.

CHAPTER 8

The Pleiotropic Effect of Physical Exercise on Mitochondrial Dynamics in Aging Skeletal Muscle

ELENA BARBIERI, DEBORAH AGOSTINI, EMANUELA POLIDORI, LUCIA POTENZA, MICHELE GUESCINI, FRANCESCO LUCERTINI, GIOSUÈ ANNIBALINI, LAURA STOCCHI, MAURO DE SANTI, AND VILBERTO STOCCHI

8.1 INTRODUCTION

Aging is associated with a generalized decline in all physiological functions, and between the ages of 30 and 70 we are likely to observe a 25–30% reduction in most functional capacities [1]. Unfortunately, this physiological condition become clinically relevant in more than 25% of people older than 85 years [2] and is referred to as "frailty." This condition has been defined as a state of increased vulnerability to poor resolution of homoeostasis after a stressor event, which increases the risk of adverse outcomes [3–5]. In other words, an apparently small insult, such as a mi-

The Pleiotropic Effect of Physical Exercise on Mitochondrial Dynamics in Aging Skeletal Muscle. © *Barbieri E, Agostini D, Polidori E, Potenza L, Guescini M, Lucertini F, Annibalini G, Stocchi L, De Santi M, and Stocchi V.* Oxidative Medicine and Cellular Longevity **2015** *(2015), http://dx.doi. org/10.1155/2015/917085. Licensed under a Creative Commons Attribution 3.0 Unported License, http://creativecommons.org/licenses/by/3.0/.*

nor infection or surgery, results in disproportionate changes in the health state. Although either the brain or the endocrine and immune system can be affected by frailty, the aging skeletal muscle has been regarded as the key component of frailty (see Clegg et al. [6]). The physiological decline of skeletal muscle function with aging, referred to as "sarcopenia," is characterized by a progressive loss of neuromuscular performance, skeletal muscle mass, and stem cell function associated with loss of strength. This intrinsic muscle weakness, also known as a deterioration in "muscle quality" has traditionally been attributed to impaired ATP production, decrease in fiber specific tension, reduced excitation-contraction coupling, and reduced neural drive [7].

Furthermore, it has been reported that adults over the age of 60 spend most of their waking hours, 8 to 12 hours per day, engaged in sedentary pursuits [8]. Inactivity accelerates muscle catabolism, mitochondrial dysfunction, and oxidative stress accumulation and reduces aerobic capacity [9]. These problems can lead to a "vicious circle" of muscle loss, injury, and inefficient repair, causing elderly people to become increasingly sedentary over time. Thus, it is imperative to implement preventive and therapeutic strategies to boost muscle mass and regeneration in the elderly and hence maintain and improve both their health and independence and prevent the occurrence of the frailty condition.

Current evidence certainly indicates that a regular exercise program reduces and/or prevents a number of functional declines associated with aging. Since, besides genetic, environmental, and nutritional factors, the lack of physical activity plays a major role in the pathophysiology of frailty [6], regular exercise has also the potential to reduce the incidence of this problematic expression of population aging. Older adults can adapt and respond to both endurance and strength training. Aerobic/endurance exercise helps to maintain and improve cardiovascular and respiratory function, whereas strength/resistance-exercise programs have been found to be helpful in improving muscle strength, power development, and function [10]. In this age group, a regular exercise program also reduces the risk factors associated with chronic disease, such cardiovascular disease, diabetes, and osteoporosis, improving overall health and helping to increase lifespan [11]. Together, these training adaptations greatly improve "muscle quality" and functional capacity on the elderly, thus improving their quality of life.

The present review aims to assess the role of exercise in enhancing mitochondrial function, biogenesis, dynamics, turnover, and quality control in aging muscle, as an area of research on bioenergetics and homeostasis, which has placed the mitochondria at the center of these processes. Exercise induces beneficial adaptations for metabolic homeostasis. This could lead to significant changes in lifestyle, which could slow down the progression of age-related muscle functional decline and could also allow us to identify molecular responses that may be useful as both therapeutic targets and for exercise prescription.

8.2 NEW EVIDENCE SUPPORTING THE MITOCHONDRIAL THEORY OF AGING AND THE ROLE OF MTDNA

Although several theories have been suggested to clarify the mechanisms mediating aging, the "Free Radical Theory," proposed in 1956 by Harman, is by far the most popular. This theory proposes that aging depends on oxidative modifications caused by highly reactive compounds such as free radicals, the most important of which are reactive oxygen species (ROS) and reactive nitrogen species (RNS) [12]. Later this theory was revised, identifying the mitochondria both as the primary sources of ROS and the primary targets of ROS damage [13, 14]. This new hypothesis, also called the "Mitochondrial Free Radical Theory of Aging" (MFRTA), is mainly based on the accumulation of ROS-mediated mtDNA damage, which arises from a "vicious circle" of ROS production, mtDNA mutations, and mitochondrial dysfunction. However, this idea still leaves many unanswered questions.

López-Otín et al. describe nine tentative hallmarks for aging in different organisms, suggesting that the rate of aging is controlled, at least to some degree, by genetic pathways and biochemical processes conserved in evolution [15]. These hallmarks are genomic instability, telomere attrition, epigenetic alterations, loss of proteostasis, deregulated nutrient sensing, mitochondrial dysfunction, cellular senescence, stem cell exhaustion, and altered intercellular communication. Nevertheless, mitochondrial dysfunction remains a common denominator of aging and it is considered to be part of the compensatory or antagonistic responses to the damage caused by cellular aging.

More recently Shokolenko has carried out a critical analysis on the role of mtDNA in aging, providing evidence that goes against the existence of the "vicious circle" [16].

MtDNA mutations observed in aging are randomly distributed and vary in type according to whether they occur in mitotic and postmitotic tissues. Point mutations are mostly observed in mitotic tissues, while large-scale deletions occur in postmitotic tissues [17]. In aging cells the highest rate of point mutations is represented by transitions, with transversions and small deletions equally distributed. Nonsynonymous and frameshift/premature termination codon point mutations have a significantly elevated frequency, as well as the predicted pathogenicity, when compared with the same variants in general populations. Since there is not a selective advantage for deleterious point mutations more than 60% of cellular copies of a given mtDNA-encoded trait have to be affected by a pathogenic mutation in order to observe the phenotypic manifestation and the impairment of mitochondrial function. These levels indeed are not achieved in naturally aged tissues of animals or humans [18–21]; thus these relatively low mutation loads, though in a heteroplasmic state, cannot be the driving force of age related decline in mitochondrial function. Hence increased ROS production in aging leads to increased mtDNA mutagenesis, though at present we do not fully understand how oxidative mtDNA lesions are processed by mitochondria to produce mutations.

Moreover it has been suggested that oxidative stress stimulates the expression of endogenous mitochondrial DNA methyltransferases (DNMTs) responsible of epigenetic modifications able to regulate the mitochondrial transcription [22] and promotes the increase of 8-hydroxy-2'-deoxyguanosine (8-oxodG) both in the nuclear and in mtDNA genome [23, 24]. Similarly in nuclear genome the increase of 8-oxodG in CpG nucleotide [25] seems to diminish the ability of DNA methyltransferases (DNMTs) to methylate the adjacent cytosine [26]. In fact, it significantly reduces the affinity of the methyl-CpG binding domain proteins (MBDs), necessary for the 5mC-mediated transcriptional repression, to their recognition sequence, thus compromising the transcription. Nevertheless whether this mechanism is also operative in mitochondria and, if so, whether it leads to the same consequences observed in the nuclear DNA requires further investigation.

According to MFRTA ROS-induced mtDNA damage should increase ROS production and trigger the "vicious" circle. Instead most of mtDNA point mutations do not affect ROS levels, as observed in "mutator-mice" [27–30] and also in humans [31]. Mito-mice are a knock-in model of mice that age prematurely due to the accumulation of random mtDNA mutations. Based on this evidence Barja has recently proposed a new version of MFRTA that does not include the "vicious circle" [32].

It is of note that the most recent reports on mtDNA damage, ROS, and aging explain the increased susceptibility of mtDNA strand breaks (but not the oxidative base damage) induced by ROS as a mitochondria-specific mechanism that helps to preserve mtDNA integrity. Thus increased ROS production has now come to be viewed as an adaptive response of mitochondria [16], often called mitohormesis, to mitigate dangerous changes rather than representing an inevitable byproduct of mitochondrial respiration. Mitohormesis also increases stress resistance, maintains mtDNA levels, preserves mtDNA fidelity, enables cells to tolerate high levels of mtDNA mutations [33], and generally prolongs lifespan [34–36]. This represents a novel-upsetting paradigm, which explains the failure of antioxidants to delay aging in clinical trials [37, 38].

Also of particular interest is the recently described circulating mtDNA that seems to increase with age. Indeed, when released extracellularly, it can act as "damage-associated molecular pattern" agent and it has been recognized as a cause of inflammation that can significantly contribute to the maintenance of the low-grade, chronic inflammation observed in elderly people [39].

At present we need to accurately determine the signature of oxidative stress in mitochondria; thus the development of reliable methods to detect the identity and the rate of ROS generated in vivo will allow us to gain further insights into the contribution of mtDNA in aging.

8.3 AGING SKELETAL MUSCLE

Human aging is associated with both loss of muscle mass and structural alteration of the neuromuscular components resulting in impaired contractile function. Sarcopenia refers to the loss of muscle mass and, consequently, strength, with aging. The sarcopenic phenotype is characterized by a reduction of muscle mass, a shift in fiber-type distribution, associated with the loss of the ability to generate force, and hence an inability to effectively perform the activities of daily living (ADL). The phenotype does not necessarily include malaise. It has recently been confirmed that differences in the leg muscle cross sectional area (CSA) between young and elderly men can be attributed to differences in type II muscle fiber size [40]. In addition, it has been shown that aging is associated with the replacement of muscle fibers with intramuscular fat and connective tissue and with oxidative stress, degeneration of the neuromuscular junction, and changes in muscle metabolism, which lead to progressive loss of muscle function and frailty. Concomitant with shift in fibre-type distribution, a shift in energy metabolism seems to anticipate the onset of sarcopenia: indeed, activity of cytochrome c oxidase that it is involved in mitochondrial oxidative phosphorylation significantly decrease and the subcellular organization of mitochondria in oxidative fibre results compromised [41] (see paragraph "Aged-related changes in skeletal muscle mitochondria").

Changes in the endocrine muscle microenvironment during aging might contribute to the progression and reduced reversibility of sarcopenia. Thus, for example, during aging production of insulin-like growth factor (IGF-I) and other anabolic molecules such as cytokines in muscle tissue declines, which could explain the reduced synthesis of myofibrillar and mitochondrial proteins with age [42]. It is known that muscle-specific IGF-I overexpression and its receptors attenuates the age-related loss of muscle mass [43] and has long been recognized as one of the critical factors for coordinating not only muscle growth, but also enhancing muscle repair and increasing muscle mass and strength [44].

IGF-I has also been found to contribute to oxidative balance and to mediate protective responses against oxidative stress in vivo [45]. IGF-I is a peptide hormone with a complex posttranscriptional regulation and generates distinct isoforms, namely, IGF-IEa, IGF-IEb, and IGF-IEc [46]. In murine models, the local muscle isoform of IGF-I (mIGF-I, the orthologue of human Mechano Growth Factor) has been shown not only to activate proliferation of myoblasts [46], but also to protect cardiomyocytes from oxidative stress via the Sirtuin 1 deacetylase activity [47]. However, chronic IGF-I supplementation might not be a healthy treatment for sarcopenia and cachexia since high level of insulin-IGF-I signaling, at least circulating, has been shown to favor cancer growth and to shorten lifespan [48].

In general, aged muscles are less responsive to anabolic and catabolic stimuli than young muscles. Indeed, menopause typically induces the reduction of about 15% of muscle mass and a decrease in estrogens levels in women, thus playing a potential role in this decline. Sarcopenia is known to be the result of unfavourable and detrimental effects, which involve hormonal, biological, nutritional, and physical activity-related mechanisms. It is challenging to determine the relative contribution of sex hormones on the sarcopenia establishment because the effect of hormonal supplementation to treat or prevent sarcopenia seems to be contradictory. On the contrary, it remains evident that the decline in muscle mass is related to an increased risk of functional impairment and physical disability [49]. Despite the rapid drop down of estrogens in women with menopause, serum testosterone levels in men decline lightly progressive starting from the third decade. Testosterone increases muscle protein synthesis and its effects on muscle are modulated by several factors, including genetic background, nutrition, and exercise. A recent review showed that testosterone supplementation produced an inferior increase of muscle CSA than resistance training alone [50]. Glucocorticoids have a reduced catabolic effect in aged sarcopenic rats and do not succeed in enhancing protein breakdown [51]. The cellular mechanisms at the base of altered sensitivity to anabolic and catabolic stimuli in aging and the physiological consequences of such reduced sensitivity to extrinsic factors remain thus unclear.

Data from the Baltimore Longitudinal Study of Aging [52] and the Health ABC study [53] showed that during aging muscle strength declines faster than muscle mass, suggesting a decrease in muscle "quality." A decrease in type II muscle fibers is specifically related to overall muscle "quality." In fact, evidence indicates that, as we age, peak force significantly decreases in these fibers and not in type I fibers [54]. Likewise, the increase in muscle mass that occurs following prolonged resistance-type exercise training (RT) can be entirely and unambiguously attributed to type II muscle fiber hypertrophy [55]. Therefore, any intervention counteracting sarcopenia should target type II muscle fiber hypertrophy by means of RT. Exercise programs, which include daily RT, should aim to prevent older, high-risk people from falling and to improve and maintain their functional capacity [56].

Both maximal rate of oxygen consumption (VO_2) and resting VO_2, when corrected for lean mass [57], decline with age in healthy individuals. Furthermore, since sedentary older people have a low arterial-venous difference, that is, their muscle oxygen extraction is reduced compared to younger adults at the maximal VO_2 rate, it follows that both muscle mitochondrial content and function are reduced in the elderly [9]. A sedentary lifestyle, typical of the elderly, accelerates skeletal muscle dysfunction, which is not only due to simple reduction in muscle mass. Intrinsic muscle weakness, also known as deterioration in "muscle quality," has traditionally been attributed to a decrease in the specific tension of the fiber and reduced neural drive. Calcium (Ca^{2+}) accumulation in energized skeletal muscle mitochondria has emerged as a biological process of great physiological relevance [58]. The reduced amount of Ca^{2+} ions available to tolerate muscle contractions and impaired ATP production due to increased muscle utilization [59] are crucial to explaining the reduced strength and resistance to fatigue of skeletal muscle in response to exercise that occurs during aging. The malfunction of excitation-contraction coupling, the mechanism linking the action potential to Ca^{2+} release from the sarcoplasmic reticulum, is probably caused by the age-related decrease in the number of calcium release units (CRUs) as described by Zampieri et al. [60]. Moreover, low ATP generation could depend on mitochondrial impairment and misplacement in the microdomain of the CRUs crucial to mitochondrial membrane potential signal activation and for efficient ATP production [61] (see paragraph "Aged-related changes in skeletal muscle mitochondria"). Given that, the molecular identity of mitochondrial Ca^{2+} uniporter (MCU) has been discovered only recently [62]; the identity of the specific complex of anchoring that stabilizes the Mitochondrial-Sarcoplasmic Reticulum is still unknown. Increased intramyocellular lipid (IMCL) content and interfiber fat infiltration is associated with aging and inactivity. Older adults have larger IMCL droplets, higher content in the subsarcolemmal area, fewer mitochondria, and a lower proportion of IMCL in contact with mitochondria compared with IMCL occurring in young skeletal muscle. These factors likely contribute to age-related reductions in mitochondrial function and oxidative metabolism [63]. Magnetic resonance spectroscopy studies have shown that IMCL levels vary with insulin sensitivity and obesity, which is a common clinical picture in the older adult, and that inactivity combined with overconsumption of fat can have detrimental effects on muscular insulin sensitivity [64]. Endurance training (ET) increases mitochondrial content/activity and IMCL content in young, active men and women. ET induces positive changes in mitochondrial function and lipid oxidation and induces intracellular IMCL reorganization, which is reflective of a greater IMCL turnover capacity in both lean and obese women [65].

In accordance, the enzyme phosphoenolpyruvate carboxykinase (PEPCK), mainly linked to gluconeogenesis, has been recently associated with a prolonged lifespan. Eukaryotes have a gene for both a mitochondrial (PEPCK-M) and cytosolic (PEPCK-C) form of the enzyme. Skeletal muscle has a small but significant level of PEPCK-C activity [66]. There have been several proposals regarding the metabolic role of this enzyme in muscle, among these we have the production of pyruvate for the synthesis of alanine by alanine aminotransferase; another possible metabolic role of PEPCK-C in skeletal muscle is glyceroneogenesis. In order to determine the metabolic role of PEPCK-C in skeletal muscle, transgenic mice were generated (PEPCK-C^{mus} mice) [67]. During exercise, the PEPCK-C^{mus} mice mainly use fatty acid as the primary fuel, the stimulation of the aerobic metabolism in the PEPCK-C^{mus} mice could be due to the increased number of mitochondria noted in their skeletal muscle [67]. Another interesting feature of the PEPCK-C^{mus} mice is that they lived almost two years longer than the controls and had normal litters of pups at 30 to 35 months of age [67]. This evidence is very interesting because these mice violate the idea that limiting food intake increases longevity, as a matter of fact, the PEPCK-C^{mus} mice eat almost twice as much as controls. Altogether, these data suggest that sustained activity could be a key element to extend lifespan and counteract sarcopenia.

In this contest, is sarcopenia reversible? Several pharmacological treatments, including selective androgen receptor molecules, anti-myostatin antibodies and activation of notch-mediated satellite cell proliferation, have been recently explored [7]. Individual physical activity history and aerobic and resistance training interventions, which are the major focus of this review, seem to be associated with the reversibility of sarcopenia.

Indeed, recent findings suggest that regular skeletal muscle contraction, such as a resistance training program of at least 12 weeks [68] or a combination of ET and RT activities [69] counteract the detrimental effects of a sedentary lifestyle, as well as the regular use of neuromuscular electrical stimulator [70]. Indeed, they represent a good intervention to attenuate and slightly reverse the decline of skeletal myofiber size, strength, and power associated with the ultrastructural disorders observed during aging. In agreement with those evidences, Melov et al. [71] highlighted that six months of resistance exercise training markedly reversed the transcriptional profile of elderly back to that of adulthood; this response was observed for most genes that are affected by both age and exercise, with a general improvement for transcripts related to mitochondrial function. Moreover a recent 4-year follow-up study on Chinese elderly people confirmed that the accomplishment of sarcopenia reversibility is associated with several lifestyle-related factors. In particular a high BMI resulted protective against sarcopenia occurrence; however, the increment of physical activity and the maintenance of a healthy weight was beneficial in the prevention of sarcopenia as well [72]. Further studies in different populations and with a longer follow-up are necessary to better investigate to what extent lifestyle behaviours might contribute to sarcopenia reversibility.

8.4 AGED-RELATED CHANGES IN SKELETAL MUSCLE MITOCHONDRIA

In the elderly, a significant proportion of the skeletal muscle mitochondria alter their ultrastructure and subcellular localization. Indeed, in older people, mitochondria in skeletal muscle appear enlarged, more rounded in shape, with matrix vacuolization and shorter cristae [73] by comparison with mitochondria found in young people. Moreover, a greater proportion of mitochondria in the elderly are depolarized or nonfunctional, which may be indicative of defects in mitochondrial turnover [74]. The density of mitochondria in skeletal muscle also drops considerably with aging [75], as shown, for example, by means of electron microscopy in the vastus lateralis muscle of people over 60 years of age when compared to their younger counterparts [60, 76].

Boncompagni et al., with an elegant study, observed a decrease in the frequency of CRUs at sarcomere's I-A band transition and tethered mitochondria [61] with aging, as mitochondria are reduced in terms of content, function, and turnover in both subsarcolemmal and intermyofibrillar pools [77]. In addition, muscle aging is associated with the progress of a segregated SR Ca^{2+} pool that uncouples from the E–C coupling machinery. The dynamic nature of Ca^{2+} sparks appears to be misplaced in aged skeletal muscle. This condition may lead to excessive mitochondrial ROS production and alteration of the cellular redox state that contributes to change Ca^{2+} release/reuptake [78, 79]. Changes in the E–C coupling apparatus and $[Ca^{2+}]i$ homeostasis may act as causal factors of, or adaptive responses to, muscle aging. Since mitochondria-CRUs cross talk seems to be crucial for efficient ATP production, impairment in ATP production may depend on mitochondrial dysfunction and possibly reduced number and misplacement. Ultrastructural data showed in the article by Zampieri et al. [70] indicate that CRU is better preserved in subjects who exercise regularly than in sedentary individuals. Mitochondrial occurrence appears higher in athletic than in sedentary seniors with parameters similar to those of healthy young subjects. Indeed a dual amelioration, that is, a mitochondrial higher frequency and an improved positioning of both organelles, is observed. Thus lifelong physical activity may counteract age-related decline of muscle functional output and muscle fibre ultrastructure. The decrease in skeletal muscle mitochondria is associated with an overall decline in mitochondrial dynamics and impairment of AMP-activated protein kinase (AMPK), which stimulates biogenesis and functionality [80]. However, the causes of this impairment are poorly understood.

How mtDNA damage, previously described as generally linked to ROS accumulation (see paragraph "New evidence supporting the mitochondrial theory of aging and the role of mtDNA"), may affect mitochondrial function is still open to debate both in the context of general cellular homeostasis and in sarcopenia [73]. Is mtDNA damage a consequence or a cause of the muscle deterioration process associated with aging? On one hand, mitochondrial energy reduction occurring before mtDNA mutations is often detectable. On the other hand, several studies in humans reveal strong correlations between mtDNA mutation rates and bioenergetic deficiency (typically complex IV) or muscle fiber atrophy [73].

Moreover, recent studies on aging skeletal muscle report that complex I and complex IV activities decrease substantially, probably because these two complexes have more of their subunits encoded by the more vulnerable mtDNA than the other complexes [73]. Age-associated mitochondrial dysfunction leads an accumulation of ROS and oxidative modification to macromolecules including proteins and failure of protein maintenance and turnover. As previously described, complexes I and III are the mitochondrial electron transport chain (ETC) sites of major ROS production. It was hypothesized that proteins of the ETC complexes are primary targets of ROS-mediated modification damaging structure and function and decline in tissue function [81]. An increased oxidative damage at the total proteome level is supposed to have a causative role in cellular aging [82], even if oxidatively modified proteins (Oxi-Proteome) have not been completely identified. In this regard, Baraibar et al. [83] have recently generated a database of proteins, which have been identified as increasingly carbonylated or modified by the lipid peroxidation product 4-hydroxy-2-nonenal (HNE) and by glycation (AGE) during aging, showing a conservation of several targeted molecules in different tissues and organ systems. Ahmed et al. [84] identified modified proteins using a proteomic approach coupled with immune-detection of HNE-, AGE-modified, and carbonylated proteins during replicative senescence of WI-38 fibroblasts. They identified by mass spectrometry thirty-seven proteins involved in quality control, energy metabolism, and cytoskeleton, showing that almost half of them were found to be mitochondria-related, underlining the susceptibility of mitochondria to senescence. A better knowledge of the Oxi-Proteome will allow to fully understanding the biological significance of these modifications [85]. However, it has become increasingly clear that most of the declines in mitochondrial biogenesis, turnover, and function are a consequence of physical inactivity. Indeed, when physical activity levels are matched between young and elderly people, or physical activity is otherwise taken into account, most investigations do not find any age-related changes in mitochondrial enzyme activities, mitochondrial respiration, or ATP flux [75, 86, 87].

To date, the endocrine mediators involved in mitochondrial (dys)function in skeletal muscle have not been deeply investigated. Recently, Puche et al. [88] demonstrated that IGF-I replacement therapy, in aging, induced mitochondrial protection, suggesting an IGF-I mediated cytoprotection effect (verified at least in neurons and hepatocytes). The authors demonstrated that the administration of low doses of IGF-I in elderly people, in which circulating IGF-I serum levels result as low as 50% compared to healthy older adults, was able to exert many beneficial effects on age related-changes, such us increasing testosterone levels, improving insulin resistance and lipid metabolism, and reducing oxidative damage on brain and liver. These benefits seem to be associated to mitochondrial protection mechanisms induced by the restore of IGF-I circulating levels. In agreement with this evidence, also Hernández-Aguilera and collaborators [89] showed that mitochondrial defects are linked to the IGF-I mTOR pathway, which is essential for the regulation of numerous processes, including cell cycle, energy metabolism, immune response, and autophagy.

Moreover, the study of steroid mediators of mitochondrial (dys)function in skeletal muscle is an emerging field and only few recent papers give some insight. For example, current studies demonstrate that mitochondria are important for the initial step of steroidogenesis, and sex steroid hormones are able to modulate mitochondrial biogenesis and function [90]. Indeed, the detection in mitochondria of glucocorticoid, steroid, and thyroid hormone receptors suggests their potential direct role in mitochondrial functional regulation [91, 92]. Dysregulation of mitochondrial function and sex steroid hormone action may compromise the maintaining cellular physiology and integrity and lead to a progressive decline in tissue function, accelerating the aging-associated phenotypes [90]. Reduction in specific endocrine regulation and accumulation of mitochondrial damage may create a feedback loop that favours the degeneration of tissue function during aging. Therefore, muscle cell changes associated with endocrine alterations and mitochondrial dysfunction require further attention.

8.5 DECREASED MITOCHONDRIAL DYNAMICS AND QUALITY CONTROL EVENTS

Mitochondria are highly dynamic organelles organized in a tubule reticulum. They constantly exchange components during biogenesis, fusion-fission events [93] which can be drastically changed by aging. In particular, the mitochondrial biogenesis signaling activated by the peroxisome proliferator-activated receptor gamma coactivator (PGC-1) family of cotranscription factors is reduced with increasing age [73]. The overexpression of PGC-1α in the skeletal muscle of aged mice improves oxidative capacity, suppresses mitochondrial degradation, and prevents muscle atrophy [94].

Building on our group's previous research in this area [95, 96], we have recently studied the role of PGC-1α in C2C12 myoblasts subjected to oxidative stress during the early stages of differentiation, as well as the effect of H_2O_2 (0.3 mM) on PGC-1α expression and its relationship with AMPK activation. We found that 1 h treatment with H_2O_2 causes mitochondrial fragmentation and a mild mitochondrial impairment with a consequent marked increase in PGC-1α mRNA expression, as described by Kang et al. [97] and Irrcher et al. [98]. In this condition, we also found an increased phosphorylation of AMPK compared to untreated cells. This suggests that oxidative stress may induce PGC-1α through the AMPK signaling pathway, probably in order to activate a defense-oriented signaling cascade (unpublished data).

However, although the defense signaling activation by AMPK phosphorylation was challenged, C2C12 myoblasts rapidly displayed a 30–40% reduction in their viability as well as a survivors' reduced differentiative efficiency during the postchallenge incubation stage (up to 7 days of culture). This observation implies that, in addition to probably being an obligatory and physiological response to ROS, activation of AMPK and of PGC-1α may not be sufficient to afford complete protection to cells against overwhelming oxidative stress. Thus increased mitochondrial production of ROS is involved at multiple levels in promoting apoptosis in skeletal muscle cells, an event which is part of the etiology and progression of numerous pathologies including sarcopenia and muscle disuse atrophy, as well as aging.

Romanello et al. [99] provided direct evidence of the importance of the existence of mitochondrial fragmentation as an amplifying circle in muscle atrophy. The mitochondrial network fragmentation induces energy, imbalance, which activates a FoxO3-dependent atrophy program through the AMPK pathway. Thus, mitochondria play a crucial role in catabolic muscle signaling: the mitochondrial fragmentation activates the AMPK-FoxO3 axis, which induces the expression of atrophy-related genes, protein breakdown, and muscle loss. The dual role of AMPK has been reviewed by Mihaylova and Shaw [100]. Authors described that activation of AMPK promotes the mitochondrial biogenesis via PGC-1α upregulation and simultaneously triggers the destruction of existing defective mitochondria through mitophagy. However, under energy stress conditions, the ROS-positive feedback loops on FoxO3 activity is acutely enhanced by AMPK sustaining autophagy and protein breakdown, both strongly related to muscle atrophy during myopathies and sarcopenia [99].

Mitochondrial fusion and fission contribute to mitochondrial function by exchanging components such as membrane, proteins, and DNA. Mitochondrial fission and fusion are regulated by GTPases of the Dynamin family, with opposite functions. Fission is mediated by dynamin related protein 1 (Drp1) and plays a key role in maintaining mitochondrial quality and mtDNA integrity, as it allows dysfunctional mitochondria to be severed from the network and to be removed by autophagy. Fusion is controlled by optic atrophy 1 (Opa1), mitofusin 1 (Mfn1), and mitofusin 2 (Mfn2 [101, 102]). Opa1 is also involved in degradative processes; indeed it regulates apoptosis by keeping the inner mitochondrial cristae junctions tight to prevent cytochrome c release, which characterizes apoptosis [103]. Few recent studies have reported that Mfn2 gene expression is lower in the skeletal muscle of older humans [63].

Of particular interest are data from muscle-specific Mfn1- and Mfn2-knockout mice. In these mice, researchers noted an enhanced mitochondrial proliferation and increased mutations in depletion of mtDNA and these changes occurred with accelerated muscle loss [33]. Age-associated changes in the dynamic remodeling processes of fission and fusion likely affect mtDNA integrity, respiratory function, ROS production, and cellular senescence.

Regarding other dynamic mitochondrial features, there is growing interest in mitochondrial quality control events in the aging of skeletal muscle.

Mitochondrial quality control involves survey, protection, and rescue strategies to limit mitochondrial damage and ensure mitochondrial integrity and function. Mitochondrial quality control involves three main steps [104]: (i) the first step that occurs at the molecular levels for the degradation of misfolded or damaged mitochondria is supported by the proteolytic system. Chaperones and ATP-dependent proteases in the matrix and inner membrane of mitochondria degrade or stabilize misfolded proteins and promote their proteolysis.

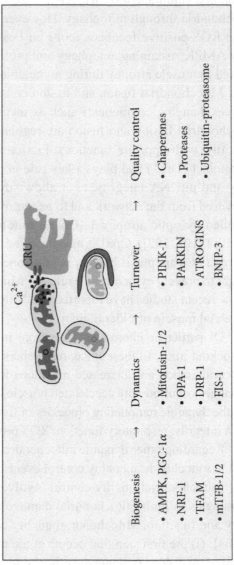

Skeletal muscle mitochondria aging and exercise training

Exercise RT/ET   Sedentary

Low levels   ROS   High levels

Mitochondrial function/mtDNA stability   Mitochondrial dysfunction/mtDNA damage
Antioxidant capacity   Mito-/apoptosis
Muscle adaptation   Sarcopenia
Gene shifting
(↑ IGF-1; muscle mass; neural activation; activated satellite cells)

Biogenesis ↑   Dynamics ↑   Turnover ↑   Quality control
• AMPK, PGC-1α   • Mitofusin-1/2   • PINK-1   • Chaperones
• NRF-1   • OPA-1   • PARKIN   • Proteases
• TFAM   • DRP-1   • ATROGINS   • Ubiquitin-proteasome
• mTFB-1/2   • FIS-1   • BNIP-3

Ca^{2+}

CRU

FIGURE 1: Effect of physical exercise and major signalling pathways activated on mitochondrial "quality" in aging skeletal muscle. Mitochondria represent the privileged site of ROS production. ROS may either act as signalling molecules, inducing a prosurvival response with positive muscle adaptation, or cause damage to cell components and sarcopenia. Low levels of ROS generated by skeletal muscle contraction activate a mitochondrial response that ameliorate the "quality" of skeletal muscle mitochondria cells at different molecular levels: (i) biogenesis through the action of the key regulators PGC-1α, NRF-1/2, T-FAM, and mTFB-1/2; (ii) dynamics by the mitochondrial remodeling GTPase proteins such as mitofusin-1/2 and OPA-1 for fusion and DRP-1 and FIS-1 for fission; (iii) turnover of damaged mitochondria by mitophagy through PINK-1, PARKIN, ATROGINS, and BNIP-3; and (iv) quality control by degradation of misfolded proteins or again portion of damaged mitochondria by the proteolytic system with chaperones and proteases. Slight ROS accumulation also promotes the phosphorylation state of many proteins involved in the muscle signalling responses. Moreover, low levels of ROS induced by RT play an important role in inducing upregulation of growth factors such as IGF-I. The expression of this muscle hormone has beneficial effects in muscle protein balance, muscle adaptation, and increasing muscle mass; neural activation; and number of activated satellite cells and contributes to the development of an oxidant-resistant phenotype, therefore preventing oxidative damage and chronic diseases. Moreover, the incorporation of satellite cell-derived mitochondria explains the increase in wild-type mtDNA known as "gene-shifting." Thus, low levels of ROS elicit positive effects on muscle physiological responses. Moreover, antioxidant enzymes will function as back regulators of intracellular low ROS levels. By contrast, high levels of ROS cause functional oxidative damages of proteins, lipids, nucleic acids, and cell components and promote signalling cascades for mitoptosis or apoptosis. For these reasons high levels of ROS act as worsening factors in muscle atrophy, sarcopenia, and aging-related muscle diseases. Uptake of calcium by mitochondria, together with ROS, control mitochondrial quality responses in skeletal muscle cells and it is tightly regulated by sarcomeric localization and muscle chronic contraction. It occurs at calcium release unit (CRU) mitochondrion contacts where microdomains of high calcium concentration are present. RT: resistance training; CRUs: calcium release units; mtDNA: mitochondrial DNA; ET: endurance training; ROS: reactive oxygen species.

Moreover, the ubiquitin-proteasome system contributes to the quality control of mitochondria; (ii) mitochondrial fusion and fission provide a second level of quality control against mitochondrial damage. Indeed, damaged mitochondria can be repaired by fusion with healthy mitochondria, which allows the contents of healthy and dysfunctional mitochondria to be mixed. Fission, by contrast, isolates mitochondria that are irreversibly damaged or not compatible for fusion leading to their elimination by autophagy; (iii) damaged mitochondria can finally be eliminated by autophagy. In particular, mitophagy selectively removes damaged mitochondria. It requires a specific labeling of damaged mitochondria and their recruitment into isolation membranes. Thus, NIP3-like protein X (NIX; also known as BNIP3L) outer mitochondrial membrane protein binds to LC3 on the isolated membranes, which mediate the capture of damaged mitochondria in autophagosomes. Once mitochondria are damaged by losing their membrane potential, PTEN-induced putative kinase protein 1 (PINK1) recruits the E3 ubiquitin ligase parkin from the cytosol to the damaged mitochondria and causes the mitochondria to become engulfed by isolation membranes that then fuse with lysosomes. If the level of damage exceeds the capacity of all three quality control pathways, damaged mitochondria can rupture, leading to the release of proapoptotic factors and cell death.

Evidence suggests that mitophagy selectively removes malfunctioning mitochondria that are depolarized or that are producing an excessive amount of ROS. In addition, suppression of autophagy results in increased ROS production, reduced oxygen consumption, and a higher level of mtDNA mutation rates. With age, autophagy declines [105], and it has been negatively correlated with oxidative damage and apoptosis, suggesting that inhibition of autophagy may contribute to muscular atrophy and dysfunction. This has been supported by studies on muscle-specific Atg7 knockout mice, which accumulate abnormal mitochondria and have lower resting oxygen consumption and increased oxidative stress and apoptosis. These conditions positively correlated with an atrophic, weak, and degenerative phenotype [106]. Moreover, recent studies indicate that enhanced autophagy may increase lifespan [107]. Mitochondrial quality control and degradation are also controlled by the ubiquitin-proteasome system, which removes damaged proteins and short-lived proteins. Evidence from studies in mammals has suggested that ubiquitin-proteasome activity also declines with age in skeletal muscle and may be one of the causes of muscular atrophy [108]. It has recently been shown that the AMPK agonist AICAR exhibits a strong and cancer-specific growth effect, which depends on the bioenergetic signature of the cells and involves upregulation of oxidative phosphorylation. In fact, sensitivity to the pharmacological activation of AMPK is higher when cells display a high proliferation rate.

Another mechanism for eliminating damaged mitochondria is mitoptosis. It has been shown that in energetic stress conditions and mitochondrial network fragmentation, damaged mitochondria cluster in the perinuclear region and become incorporated into a single-membraned mitoptotic body and finally undergo extrusion via exocytosis or blebbing [109]. This observation is in agreement with the characterization of microvesicles carrying mtDNA released by C2C12 recently described by Guescini et al. [110].

Maintaining well-functioning skeletal muscle mitochondrial dynamics in terms of content, function, and turnover is important for maintaining good health throughout our lives. Exercise stimulates key stress signals that regulate the skeletal muscle quality of mitochondria during aging (see Figure 1 for a summary). Perturbations in mitochondrial content and function can directly or indirectly impact skeletal muscle function and, consequently, the health of our whole body and overall well-being.

8.6 EXERCISE AND PHYSICAL ACTIVITY FOR OLDER ADULTS

It has become increasingly clear that most of the declines in skeletal muscle function attributed to chronological age are instead a result of physical inactivity. Most physical activity guidelines for apparently healthy adults recommend performing about 30 minutes of moderate intensity aerobic exercise at least 5 times per week [10]. It has also emerged that, on a given day, these 30 minutes may be accumulated in several short bouts (of at least 10 min each) distributed throughout the day rather than in one continuous bout, without reducing training efficacy. This is a noteworthy consideration because exercising for short periods may be more pleasant and may better address common barriers to physical activity such as a lack of time or facilities, high costs, poor weather conditions, and embarrassment; hence it may improve long-term adherence to this lifestyle change. Although the minimum dose necessary to improve health status remains unclear, recently a significant improvement in cardiorespiratory fitness was reported by Mair et al. after only 4 weeks [111]. In the study, previously sedentary middle-aged adults performed 30 min per week of the bench step exercise accumulated in short bouts throughout the day. This suggests that a much shorter time than is currently recommended may be sufficient to initiate improvements in fitness.

Although ET exercise is the more common method for improving cardiovascular fitness, RT is currently recommended for the elderly, in whom loss of muscle mass and weakness are major problems. Even two training sessions a week of RT are sufficient to induce an increase in muscle strength and power [69] as a result of both an increase in muscle mass and/ or in the level of neural activation.

With a similar training stimulus, the hypertrophic response seems to be blunted in older adults compared to their younger counterparts, and this phenomenon appears to be more evident in female subjects. In this regard, the role of diet, especially in terms of protein intake, remains unclear in determining muscle adaptations. In fact, it is well known that RT in elderly populations increases both mixed-muscle and MHC-specific protein synthesis to the same extent that it does in young subjects [112]. Furthermore, an increase in muscle fractional synthetic rate has been observed following RT even in frail elderly persons (>70 years) [112, 113]. RT in older adults significantly increases type II fibers CSA [54, 114, 115] as well as the proportion of type IIa fiber distribution [116].

Increasing muscle protein synthesis and muscle fiber hypertrophy corresponds to an increase in force-generating capacity and an improvement in ADL performance, leading to a significant improvement in quality of life.

RT in the elderly has also been related to decreased morbidity and even mortality [69]. The improvement in muscular strength and power induced by RT and the central and peripheral adaptations that improve cardiovascular fitness and increase energy production via oxidative metabolism induced by endurance training may be combined (i.e., concurrent training) in an efficient exercise program for the elderly that enhances both neuromuscular and cardiorespiratory functions while preserving functional ability [117]. Several studies have examined concurrent training effects on young subjects, while only few authors have investigated concurrent training adaptations in the elderly [117]. Hence, for a correct exercise prescription for the elderly it seems appropriate to identify in both exercise types the most effective combination of training variables (i.e., weekly frequency, intensity, duration, volume, exercise-order, etc.), which stimulate neuromuscular and cardiovascular adaptations to the greatest extent.

8.7 EXERCISE TRAINING AND MITOCHONDRIA "QUALITY" IN THE ELDERLY

Mitochondria play a central role in cellular metabolism and form, by fusion and fission, a dynamic reticular network within mammalian skeletal muscle [118–120]. This network allows them to share mtDNA and also to degrade and remove damaged components, through a process called mitophagy. Thus, mitochondrial biogenesis and mitophagy constantly control mitochondrial content in terms of quantity and quality [121]. Aging affects fission and fusion, mtDNA integrity, respiratory function, ROS production, cellular senescence, and also, to some extent, mitophagy. Aging also induces an increase in apoptotic cells, mostly in type II fibers [122], which contribute to sarcopenia [123].

It is well known that older adults may benefit from exercise-induced adaptations in mitochondrial biogenesis and cellular antioxidant defense [124] and that these adaptations are quickly reversed following a reduction or the cessation of physical activity.

Another important aspect to evaluate in order to understand the "trainability" of mitochondrial health and content is genetics, since there is great variability in training-induced changes in aerobic characteristics, due to nuclear genetic variants of genes [120] encoding for mt proteins and to mtDNA haplogroups [125]. The Health, Risk Factors, Training and Genetics (HERITAGE) Family Study suggests that the heritability of changes in maximal VO_2 with exercise training is ~47% in sedentary subjects [126] while older studies in twins reported a heritability of 93.4% [127], underlining the influence of the genome on exercise-induced results.

Furthermore, studies analyzing changes in mitochondrial content and quality should also take into account fiber type and the specific mitochondrial population (i.e., SS or IMM).

As mitochondria are highly sensitivity to contractile-initiated signals, physical activity, and exercise promote biogenesis and function in these organelles, helping to maintain cellular and whole body health. In any case, many factors should be taken into consideration in determining the best exercise strategy (in terms of intensity, volume, and dosage) to improve mitochondrial function. Several studies have reported an increase in skeletal muscle mitochondrial biogenesis following ET and it appears that training volume, rather than training intensity, may be an important determinant of exercise-induced improvements in mitochondrial content [128].

RT training has been shown to counteract the deleterious effect of sarcopenia [71] and enhance mitochondrial function in aging muscle [129], but it is unclear whether the cellular adaptations to RT directly counteract the process of mitochondrial aging, interfering with key aspects of the free-radical induced aging process, or simply mask it. Since older adults rarely perform extremely high-intensity and low-repetition protocols, it is possible that mitochondrial biogenesis simply occurs in response to the volume of exercise. In any case, this issue warrants further exploration.

Wilkinson et al. [130] showed that, in young adults, a single bout of RT causes both myofibrillar and mitochondrial protein synthesis before training, while, after training, it causes only myofibrillar synthesis. Tang et al. [131] also reported an increase in mitochondrial enzyme activity following RT in young adults. In older adults, Parise et al. [114, 132] showed that RT results in high levels of antioxidant enzymes and lower oxidative stress, while Tarnopolsky [129] investigating the effects of 14 weeks of RT on oxidant status and ETC did not observe changes in the activity of complexes I+III and II+III, but they observed an upregulation of complex IV. Complex IV is the terminal electron acceptor in the ETC, and an increase in its enzymatic capacity may reduce electron leakage, leading to lower ROS production [133]. It should also be noted that complex IV possesses more proteins encoded by mtDNA than other complexes and thus is more affected by mtDNA mutations. In the skeletal muscle of older adults, as well as of some patients with mitochondrial diseases, there is heteroplasmy, which means the presence of mtDNA copies belonging to wild-type or mutant mtDNA populations. The degree of mutant mtDNA heteroplasmy found in mature skeletal muscle fibers is higher than that which is found in peripheral blood mononuclear cells, fibroblasts, and satellite cells.

Although systemic mitochondrial dysfunction plays important roles in age-related muscle wasting by preferentially affecting the quiescent myosatellite cell pool [134], satellite cells are well protected against oxidative stress. Their metabolic status is important in limiting the production of ROS [135] since they have fewer mitochondria and a glycolitic metabolism, compared to oxidative activated cells. Furthermore, quiescent satellite cells overexpress genes specifically involved in controlling the cytotoxicity of ROS and they repair DNA mutation/s much more efficiently than differentiated myoblasts. Although the repair efficiency varies slightly in different experimental conditions, satellite cells systematically repair DNA damage more efficiently than their progeny under similar conditions [136].

A process referred to as "gene shifting" can contribute to explain the benefits of RT training in improving skeletal muscle mitochondrial health. Taivassalo and colleagues [137] examined the mitochondrial genotype in mature myofibers of patients with mitochondrial disease, following either an eccentric or concentric RT intervention. They found a significant increase in the amount of wild-type mtDNA, together with a dramatic decrease in the proportion of COX-negative muscle fibers. RT causes muscle overload or injury, resulting in myofiber hypertrophy or repair processes depending on activated satellite cells that can fuse to form regenerating myofibers [138]. The incorporation of satellite cell-derived mitochondria explains the increase in wild-type mtDNA known as "gene-shifting." More recently, Murphy and colleagues [139] showed that, in adults with large-scale mtDNA deletions, 12 weeks of RT induced an increase in muscle strength, myofiber damage and regeneration, NCAM-positive and COX-positive satellite cells, and oxidative capacity, supporting the RTe-induced mitochondrial gene-shifting hypothesis. Recently, Spendiff et al. [140], in a study on mitochondrial DNA deletion demonstrated that in patients with mitochondrial myopathies, satellite cells presented single, large-scale mtDNA deletion/s at levels comparable with those observed in muscle. According to them, whether mtDNA mutation/s occur in satellite cells, they are subsequently lost during satellite cell activation and myoblast proliferation. Thus, the gene shifting strategy induced by resistance exercise, although with a different mechanism from that originally proposed, is likely to be a suitable intervention in patients with myopathies.

This evidence supports the hypothesis that also in older adults, following RT, there is not only a mitophagic process able to remove dysfunctional mitochondria, but also a potential mtDNA-shifting. Further studies are needed to quantify the extent of this phenomenon in counteracting aging through RT and to identify the intensity of strength exercise required to activate satellite cells. This is particularly important to improve DNA-shifting and oxidative capacity, since it is known that training intensity is an important determinant in mitochondrial function and has less of an effect on mitochondrial content, which is more strongly related to training volume.

Due to the peculiar properties of ET and RT in promoting mitochondrial quantity and functioning quality, and their differing effects on biogenesis and genotype, different physical activity modalities should be combined to maximize the antiaging effects of exercise.

Thus, a preliminary period of ST which induces satellite cell activation and transfer of normal mitochondrial genes to existing muscle (decreasing the amount of COX deficient fibers and increasing the levels of wild-type compared to mutant), followed by ET induced mitochondrial biogenesis, which expands newly incorporated wild-type genomes, could be the most effective approach as suggested by [139].

According to the current paradigm, concurrent training, that is, endurance and resistance exercise combined in the same training session, results in a blunted response, due to interference between the different types of exercise. It is known that muscle growth is mediated by mTORC1, while mitochondrial biogenesis is driven by PGC-1α, and these two pathways are linked to one another. However, in addition to the studies supporting the interference theory, other recent studies have yielded conflicting results. Wang et al. [141] reported that endurance exercise followed by resistance exercise improves PGC-1α and activates mTORC1 also independently form IGF-I upregulation, its receptors. Apró et al. [142], who examined whether endurance exercise following heavy resistance exercise would repress molecular signaling through the mTORC1 pathway when compared with resistance exercise alone, did not find any interference with the growth-related signaling through the mTORC1 pathway in human skeletal muscle. In addition, both modes of exercise induced similar responses at the transcriptional level with the exception of PGC-1α, whose gene expression was superior following concurrent exercise. Furthermore, MacNeil et al. [143] reported that the order of exercise modes within concurrent training (endurance following resistance exercise or vice versa) does not affect training-induced changes in gene expression, protein content or measures of strength and aerobic capacity. Although it appears that concurrent training, regardless of the exercise mode order, is probably the most effective strategy for improving mitochondrial health and biogenesis, further studies are needed to identify the most useful exercise strategy in terms of intensity, volume, and timetable, also taking into account the overall physical condition of the subject. In this regard, the introduction of the minimally invasive technique called "fine-needle aspiration" has proved to be very useful in the study of mitochondrial quality and quantity modulation in response to exercise and may help in developing personalized exercise training programs particularly in elderly [144–146].

In addition, in subjects with an altered mitochondrial condition, extended muscle wasting and dysfunction, or nutraceutical interventions, not taken into consideration in this review, should be evaluated, alone or together with exercise. Creatine, for example, which has been shown to enhance satellite cell activation during resistance exercise training [147], could promote gene shifting. Moreover, encouraging results described in a recent meta-analysis by [148] support a role for Creatine (Cr) supplementation during RT in healthy elderly subjects. The RT with Cr supplementation enhanced muscle mass gain, strength, and functional performance, more than RT alone.

8.8 CONCLUDING REMARKS

In this review we describe the pleiotropic effect of physical activity on multiple targets that have a role in preventing the decline of mitochondrial "quality," which is implicated in the aging process of skeletal muscle. Recent evidences consistently show that the "quality" of skeletal muscle mitochondria declines during aging. Indeed, in this condition we can observe (i) mitochondrial DNA mutations; (ii) specific epigenetic drift; (iii) decreased expression of mitochondrial proteins; (iv) reduced enzyme activity of cellular respiration; (v) reduced total mitochondrial content; (vi) increased morphological changes; (vii) a decrease in mitochondrial turnover. All of these factors probably contribute to age-associated sarcopenia, and a growing body of evidence suggests that most of these skeletal muscle age-related changes can be prevented and or attenuated by physical activity.

The current ACSM recommendations assume that older adults can safely participate in regular exercise programs (aerobic and strength). Any given exercise program for the elderly will depend on existing co-morbidities and on the baseline level of physical activity. Elderly individuals can play an active role in designing their own exercise programs, which should aim to improve aerobic capacity, strength, and balance, all vital to healthy aging.

In short, physical activity should be prescribed for older adults. It not only improves physical function, helping the elderly to maintain independence, but also enhances overall health and increases longevity.

REFERENCES

1. G. A. Power, B. H. Dalton, and C. L. Rice, "Human neuromuscular structure and function in old age: a brief review," Journal of Sport and Health Science, vol. 2, no. 4, pp. 215–226, 2013.
2. X. Song, A. Mitnitski, and K. Rockwood, "Prevalence and 10-year outcomes of frailty in older adults in relation to deficit accumulation," Journal of the American Geriatrics Society, vol. 58, no. 4, pp. 681–687, 2010.
3. E. M. P. Eeles, S. V. White, S. M. O'Mahony, A. J. Bayer, and R. E. Hubbard, "The impact of frailty and delirium on mortality in older inpatients," Age and Ageing, vol. 41, no. 3, Article ID afs021, pp. 412–416, 2012.
4. L. P. Fried, C. M. Tangen, J. Walston et al., "Frailty in older adults: evidence for a phenotype," The Journals of Gerontology. Series A, Biological Sciences and Medical Sciences, vol. 56, no. 3, pp. M146–M156, 2001.
5. J. Walston, E. C. Hadley, L. Ferrucci et al., "Research agenda for frailty in older adults: toward a better understanding of physiology and etiology: summary from the American Geriatrics Society/National Institute on Aging Research Conference on Frailty in Older Adults," Journal of the American Geriatrics Society, vol. 54, no. 6, pp. 991–1001, 2006.
6. A. Clegg, J. Young, S. Iliffe, M. O. Rikkert, and K. Rockwood, "Frailty in elderly people," The Lancet, vol. 381, no. 9868, pp. 752–762, 2013.
7. M. V. Narici and N. Maffulli, "Sarcopenia: characteristics, mechanisms and functional significance," British Medical Bulletin, vol. 95, no. 1, pp. 139–159, 2010.
8. L. F. M. De Rezende, J. P. Rey-López, V. K. R. Matsudo, and O. D. C. Luiz, "Sedentary behavior and health outcomes among older adults: a systematic review," BMC Public Health, vol. 14, no. 1, article 333, 2014.

9. M. L. Johnson, M. M. Robinson, and S. K. Nair, "Skeletal muscle aging and the mitochondrion," Trends in Endocrinology and Metabolism, vol. 24, no. 5, pp. 247–256, 2013.

10. C. E. Garber, B. Blissmer, M. R. Deschenes et al., "American College of Sports Medicine position stand. Quantity and quality of exercise for developing and maintaining cardiorespiratory, musculoskeletal, and neuromotor fitness in apparently healthy adults: guidance for prescribing exercise," Medicine and Science in Sports and Exercise, vol. 43, no. 7, pp. 1334–1359, 2011.

11. D. E. Warburton, C. W. Nicoland, and S. S. Bredin, "Health benefits of physical activity: the evidence," Canadian Medical Association Journal, vol. 174, no. 6, pp. 801–809, 2006.

12. D. Harman, "Aging: a theory based on free radical and radiation chemistry," Journal of Gerontology, vol. 11, no. 3, pp. 298–300, 1956.

13. D. Harman, "The biologic clock: the mitochondria?" Journal of the American Geriatrics Society, vol. 20, no. 4, pp. 145–147, 1972.

14. J. Miquel, A. C. Economos, J. Fleming, and J. E. Johnson Jr., "Mitochondrial role in cell aging," Experimental Gerontology, vol. 15, no. 6, pp. 575–591, 1980.

15. C. López-Otín, M. A. Blasco, L. Partridge, M. Serrano, and G. Kroemer, "The hallmarks of aging," Cell, vol. 153, no. 6, pp. 1194–1217, 2013.

16. I. N. Shokolenko, "Aging: a mitochondrial DNA perspective, critical analysis and an update," World Journal of Experimental Medicine, vol. 4, no. 4, pp. 46–57, 2014.

17. L. C. Greaves, J. L. Elson, M. Nooteboom et al., "Comparison of mitochondrial mutation spectra in ageing human colonic epithelium and disease: absence of evidence for purifying selection in somatic mitochondrial DNA point mutations," PLoS Genetics, vol. 8, no. 11, Article ID e1003082, 2012.

18. X. Song, J. H. Deng, C. J. Liu, and Y. Bai, "Specific point mutations may not accumulate with aging in the mouse mitochondrial DNA control region," Gene, vol. 350, no. 2, pp. 193–199, 2005.

19. K. Khrapko, Y. Kraytsberg, A. D. N. J. de Grey, J. Vijg, and E. A. Schon, "Does premature aging of the mtDNA mutator mouse prove that mtDNA mutations are involved in natural aging?" Aging Cell, vol. 5, no. 3, pp. 279–282, 2006.

20. I. Shokolenko, N. Venediktova, A. Bochkareva, G. I. Wilson, and M. F. Alexeyev, "Oxidative stress induces degradation of mitochondrial DNA," Nucleic Acids Research, vol. 37, no. 8, pp. 2539–2548, 2009.

21. S. R. Kennedy, J. J. Salk, M. W. Schmitt, and L. A. Loeb, "Ultra-sensitive sequencing reveals an age-related increase in somatic mitochondrial mutations that are inconsistent with oxidative damage," PLoS Genetics, vol. 9, no. 9, Article ID e1003794, 2013.

22. L. S. Shock, P. V. Thakkar, E. J. Peterson, R. G. Moran, and S. M. Taylor, "DNA methyltransferase 1, cytosine methylation, and cytosine hydroxymethylation in mammalian mitochondria," Proceedings of the National Academy of Sciences of the United States of America, vol. 108, no. 9, pp. 3630–3635, 2011.

23. A. Valavanidis, T. Vlachogianni, and C. Fiotakis, "8-Hydroxy-2′-deoxyguanosine (8-OHdG): a critical biomarker of oxidative stress and carcinogenesis," Journal of Environmental Science and Health, Part C: Environmental Carcinogenesis and Ecotoxicology Reviews, vol. 27, no. 2, pp. 120–139, 2009.

24. L. Potenza, C. Calcabrini, R. De Bellis et al., "Effect of surgical stress on nuclear and mitochondrial DNA from healthy sections of colon and rectum of patients with colorectal cancer," Journal of Biosciences, vol. 36, no. 2, pp. 243–251, 2011.

25. P. W. Turk, A. Laayoun, S. S. Smith, and S. A. Weitzman, "DNA adduct 8-hydroxyl-2′-deoxyguanosine (8-hydroxyguanine) affects function of human DNA methyltransferase," Carcinogenesis, vol. 16, no. 5, pp. 1253–1255, 1995.

26. Y.-H. Wei, Y.-S. Ma, H.-C. Lee, C.-F. Lee, and C.-Y. Lu, "Mitochondrial theory of aging matures—roles of mtDNA mutation and oxidative stress in human aging," Chinese Medical Journal, vol. 64, no. 5, pp. 259–270, 2001.

27. C. C. Kujoth, A. Hiona, T. D. Pugh et al., "Medicine: Mitochondrial DNA mutations, oxidative stress, and apoptosis in mammalian aging," Science, vol. 309, no. 5733, pp. 481–484, 2005.

28. A. Trifunovic, A. Hansson, A. Wredenberg et al., "Somatic mtDNA mutations cause aging phenotypes without affecting reactive oxygen species production," Proceedings of the National Academy of Sciences of the United States of America, vol. 102, no. 50, pp. 17993–17998, 2005.

29. A. Trifunovic, A. Wredenberg, M. Falkenberg et al., "Premature ageing in mice expressing defective mitochondrial DNA polymerase," Nature, vol. 429, no. 6990, pp. 417–423, 2004.

30. D. Edgar, I. Shabalina, Y. Camara et al., "Random point mutations with major effects on protein-coding genes are the driving force behind premature aging in mtDNA mutator mice," Cell Metabolism, vol. 10, no. 2, pp. 131–138, 2009.

31. E. Hütter, M. Skovbro, B. Lener et al., "Oxidative stress and mitochondrial impairment can be separated from lipofuscin accumulation in aged human skeletal muscle," Aging Cell, vol. 6, no. 2, pp. 245–256, 2007.

32. G. Barja, "Updating the mitochondrial free radical theory of aging: an integrated view, key aspects, and confounding concepts," Antioxidants & Redox Signaling, vol. 19, no. 12, pp. 1420–1445, 2013.

33. H. Chen, M. Vermulst, Y. E. Wang et al., "Mitochondrial fusion is required for mtdna stability in skeletal muscle and tolerance of mtDNA mutations," Cell, vol. 141, no. 2, pp. 280–289, 2010.

34. M. Ristow, "Unraveling the truth about antioxidants: mitohormesis explains ROS-induced health benefits," Nature Medicine, vol. 20, no. 7, pp. 709–711, 2014.

35. M. Ristow and K. Zarse, "How increased oxidative stress promotes longevity and metabolic health: the concept of mitochondrial hormesis (mitohormesis)," Experimental Gerontology, vol. 45, no. 6, pp. 410–418, 2010.

36. M. Ristow, K. Zarse, A. Oberbach et al., "Antioxidants prevent health-promoting effects of physical exercise in humans," Proceedings of the National Academy of Sciences of the United States of America, vol. 106, no. 21, pp. 8665–8670, 2009.

37. R. M. Howes, "The free radical fantasy: a panoply of paradoxes," Annals of the New York Academy of Sciences, vol. 1067, no. 1, pp. 22–26, 2006.

38. G. Bjelakovic, D. Nikolova, and C. Gluud, "Antioxidant supplements to prevent mortality," Journal of the American Medical Association, vol. 310, no. 11, pp. 1178–1179, 2013.

39. M. Pinti, E. Cevenini, M. Nasi et al., "Circulating mitochondrial DNA increases with age and is a familiar trait: implications for 'inflamm-aging'," European Journal of Immunology, vol. 44, no. 5, pp. 1552–1562, 2014.
40. R. Nilwik, T. Snijders, M. Leenders et al., "The decline in skeletal muscle mass with aging is mainly attributed to a reduction in type II muscle fiber size," Experimental Gerontology, vol. 48, no. 5, pp. 492–498, 2013.
41. R. A. McGregor, D. Cameron-Smith, and S. D. Poppitt, "It is not just muscle mass: a review of muscle quality, composition and metabolism during ageing as determinants of muscle function and mobility in later life," Longevity & Healthspan, vol. 3, no. 1, article 9, 2014.
42. K. S. Nair, "Aging muscle," The American Journal of Clinical Nutrition, vol. 81, no. 5, pp. 953–963, 2005.
43. A. Musarò, K. McCullagh, A. Paul et al., "Localized Igf-1 transgene expression sustains hypertrophy and regeneration in senescent skeletal muscle," Nature Genetics, vol. 27, no. 2, pp. 195–200, 2001.
44. A. Philippou and E. R. Barton, "Optimizing IGF-I for skeletal muscle therapeutics," Growth Hormone & IGF Research, vol. 24, no. 5, pp. 157–163, 2014.
45. A. Kokoszko, J. Dabrowski, A. Lewiński, and M. Karbownik-Lewińska, "Protective effects of GH and IGF-I against iron-induced lipid peroxidation in vivo," Experimental and Toxicologic Pathology, vol. 60, no. 6, pp. 453–458, 2008.
46. B. K. Brisson and E. R. Barton, "New modulators for IGF-I activity within IGF-I processing products," Frontiers in Endocrinology (Lausanne), vol. 4, article 42, 2013.
47. A. Satoh, C. S. Brace, G. Ben-Josef et al., "SIRT1 promotes the central adaptive response to diet restriction through activation of the dorsomedial and lateral nuclei of the hypothalamus," The Journal of Neuroscience, vol. 30, no. 30, pp. 10220–10232, 2010.
48. C. J. Kenyon, "The genetics of ageing," Nature, vol. 464, no. 7288, pp. 504–512, 2010.
49. V. Messier, R. Rabasa-Lhoret, S. Barbat-Artigas, B. Elisha, A. D. Karelis, and M. Aubertin-Leheudre, "Menopause and sarcopenia: a potential role for sex hormones," Maturitas, vol. 68, no. 4, pp. 331–336, 2011.
50. K. Sakuma and A. Yamaguchi, "Sarcopenia and age-related endocrine function," International Journal of Endocrinology, vol. 2012, Article ID 127362, 10 pages, 2012.
51. M. Altun, H. C. Besche, H. S. Overkleeft et al., "Muscle wasting in aged, sarcopenic rats is associated with enhanced activity of the ubiquitin proteasome pathway," The Journal of Biological Chemistry, vol. 285, no. 51, pp. 39597–39608, 2010.
52. E. J. Metter, N. Lynch, R. Conwit, R. Lindle, J. Tobin, and B. Hurley, "Muscle quality and age: cross-sectional and longitudinal comparisons," The Journals of Gerontology—Series A Biological Sciences and Medical Sciences, vol. 54, no. 5, pp. B207–B218, 1999.
53. B. H. Goodpaster, S. W. Park, T. B. Harris et al., "The loss of skeletal muscle strength, mass, and quality in older adults: the health, aging and body composition study," The Journals of Gerontology, Series A: Biological Sciences and Medical Sciences, vol. 61, no. 10, pp. 1059–1064, 2006.

54. S. Trappe, P. Gallagher, M. Harber, J. Carrithers, J. Fluckey, and T. Trappe, "Single muscle fibre contractile properties in young and old men and women," The Journal of Physiology, vol. 552, no. 1, pp. 47–58, 2003.
55. T. van Wessel, A. de Haan, W. J. van der Laarse, and R. T. Jaspers, "The muscle fiber type-fiber size paradox: hypertrophy or oxidative metabolism?" European Journal of Applied Physiology, vol. 110, no. 4, pp. 665–694, 2010.
56. L. Clemson, M. A. Fiatarone Singh, A. Bundy et al., "Integration of balance and strength training into daily life activity to reduce rate of falls in older people (the LiFE study): randomised parallel trial," British Medical Journal, vol. 345, no. 7870, Article ID e4547, 2012.
57. D. N. Proctor and M. J. Joyner, "Skeletal muscle mass and the reduction of VO-2max in trained older subjects," Journal of Applied Physiology, vol. 82, no. 5, pp. 1411–1415, 1997.
58. C. Mammucari and R. Rizzuto, "Signaling pathways in mitochondrial dysfunction and aging," Mechanisms of Ageing and Development, vol. 131, no. 7-8, pp. 536–543, 2010.
59. M. K. Shigenaga, T. M. Hagen, and B. N. Ames, "Oxidative damage and mitochondrial decay in aging," Proceedings of the National Academy of Sciences of the United States of America, vol. 91, no. 23, pp. 10771–10778, 1994.
60. S. Zampieri, L. Pietrangelo, S. Loefler et al., "Lifelong physical exercise delays age-associated skeletal muscle decline," The Journals of Gerontology, Series A: Biological Sciences and Medical Sciences, vol. 70, no. 2, pp. 163–173, 2015.
61. S. Boncompagni, A. E. Rossi, M. Micaroni et al., "Mitochondria are linked to calcium stores in striated muscle by developmentally regulated tethering structures," Molecular Biology of the Cell, vol. 20, no. 3, pp. 1058–1067, 2009.
62. D. de Stefani, A. Raffaello, E. Teardo, I. Szabó, and R. Rizzuto, "A forty-kilodalton protein of the inner membrane is the mitochondrial calcium uniporter," Nature, vol. 476, no. 7360, pp. 336–340, 2011.
63. J. D. Crane, M. C. Devries, A. Safdar, M. J. Hamadeh, and M. A. Tarnopolsky, "The effect of aging on human skeletal muscle mitochondrial and intramyocellular lipid ultrastructure," The Journals of Gerontology Series A: Biological Sciences and Medical Sciences, vol. 65, no. 2, pp. 119–128, 2010.
64. V. B. Schrauwen-Hinderling, M. K. C. Hesselink, P. Schrauwen, and M. E. Kooi, "Intramyocellular lipid content in human skeletal muscle," Obesity, vol. 14, no. 3, pp. 357–367, 2006.
65. M. C. Devries, I. A. Samjoo, M. J. Hamadeh et al., "Endurance training modulates intramyocellular lipid compartmentalization and morphology in skeletal muscle of lean and obese women," The Journal of Clinical Endocrinology & Metabolism, vol. 98, no. 12, pp. 4852–4862, 2013.
66. E. A. Newsholme and T. Williams, "The role of phosphoenolpyruvate carboxykinase in amino acid metabolism in muscle," The Biochemical Journal, vol. 176, no. 2, pp. 623–626, 1978.
67. R. W. Hanson and P. Hakimi, "Born to run; the story of the PEPCK-Cmus mouse," Biochimie, vol. 90, no. 6, pp. 838–842, 2008.

68. S. D. Harridge, A. Kryger, and A. Stensgaard, "Knee extensor strength, activation, and size in very elderly people following strength training," Muscle & Nerve, vol. 22, no. 7, pp. 831–839, 1999.

69. American College of Sports Medicine, W. J. Chodzko-Zajko, D. N. Proctor, et al., "American College of Sports Medicine position stand. Exercise and physical activity for older adults," Medicine and Science in Sports and Exercise, vol. 41, no. 7, pp. 1510–1530, 2009.

70. S. Zampieri, L. Pietrangelo, S. Loefler et al., "Lifelong physical exercise delays age-associated skeletal muscle decline," The Journals of Gerontology. Series A: Biological Sciences and Medical Sciences, vol. 70, no. 2, pp. 163–173, 2015.

71. S. Melov, M. A. Tamopolsky, K. Bechman, K. Felkey, and A. Hubbard, "Resistance exercise reverses aging in human skeletal muscle," PLoS ONE, vol. 2, no. 5, article e465, 2007.

72. R. Yu, M. Wong, J. Leung, J. Lee, T. W. Auyeung, and J. Woo, "Incidence, reversibility, risk factors and the protective effect of high body mass index against sarcopenia in community-dwelling older Chinese adults," Geriatrics & Gerontology International, vol. 14, supplement 1, pp. 15–28, 2014.

73. C. M. Peterson, D. L. Johannsen, and E. Ravussin, "Skeletal muscle mitochondria and aging: a review," Journal of Aging Research, vol. 2012, Article ID 194821, 20 pages, 2012.

74. A. Terman, T. Kurz, M. Navratil, E. A. Arriaga, and U. T. Brunk, "Mitochondrial Turnover and aging of long-lived postmitotic cells: the mitochondrial-lysosomal axis theory of aging," Antioxidants and Redox Signaling, vol. 12, no. 4, pp. 503–535, 2010.

75. B. F. Miller, M. M. Robinson, M. D. Bruss, M. Hellerstein, and K. L. Hamilton, "A comprehensive assessment of mitochondrial protein synthesis and cellular proliferation with age and caloric restriction," Aging Cell, vol. 11, no. 1, pp. 150–161, 2012.

76. K. E. Conley, S. A. Jubrias, and P. C. Esselman, "Oxidative capacity and ageing in human muscle," The Journal of Physiology, vol. 526, no. 1, pp. 203–210, 2000.

77. B. Chabi, B. Mousson de Camaret, A. Chevrollier, S. Boisgard, and G. Stepien, "Random mtDNA deletions and functional consequence in aged human skeletal muscle," Biochemical and Biophysical Research Communications, vol. 332, no. 2, pp. 542–549, 2005. A. E. Rossi, S. Boncompagni, and R. T. Dirksen, "Sarcoplasmic reticulum-mitochondrial symbiosis: bidirectional signaling in skeletal muscle," Exercise and Sport Sciences Reviews, vol. 37, no. 1, pp. 29–35, 2009.

78. N. Weisleder, M. Brotto, S. Komazaki et al., "Muscle aging is associated with compromised Ca2+ spark signaling and segregated intracellular Ca2+ release," The Journal of Cell Biology, vol. 174, no. 5, pp. 639–645, 2006.

79. R. M. Reznick, H. Zong, J. Li et al., "Aging-associated reductions in AMP-activated protein kinase activity and mitochondrial biogenesis," Cell Metabolism, vol. 5, no. 2, pp. 151–156, 2007.

80. K. B. Choksi, J. E. Nuss, J. H. DeFord, and J. Papaconstantinou, "Age-related alterations in oxidatively damaged proteins of mouse skeletal muscle mitochondrial electron transport chain complexes," Free Radical Biology & Medicine, vol. 45, no. 6, pp. 826–838, 2008.

81. B. S. Berlett and E. R. Stadtman, "Protein oxidation in aging, disease, and oxidative stress," The Journal of Biological Chemistry, vol. 272, no. 33, pp. 20313–20316, 1997.

82. M. A. Baraibar, L. Liu, E. K. Ahmed, and B. Friguet, "Protein oxidative damage at the crossroads of cellular senescence, aging, and age-related diseases," Oxidative Medicine and Cellular Longevity, vol. 2012, Article ID 919832, 8 pages, 2012.

83. E. K. Ahmed, A. Rogowska-Wrzesinska, P. Roepstorff, A.-L. Bulteau, and B. Friguet, "Protein modification and replicative senescence of WI-38 human embryonic fibroblasts," Aging Cell, vol. 9, no. 2, pp. 252–272, 2010.

84. M. A. Baraibar, M. Gueugneau, S. Duguez, G. Butler-Browne, D. Bechet, and B. Friguet, "Expression and modification proteomics during skeletal muscle ageing," Biogerontology, vol. 14, no. 3, pp. 339–352, 2013.

85. A. Safdar, M. J. Hamadeh, J. J. Kaczor, S. Raha, J. deBeer, and M. A. Tarnopolsky, "Aberrant mitochondrial homeostasis in the skeletal muscle of sedentary older adults," PLoS ONE, vol. 5, no. 5, Article ID e10778, 2010.

86. I. R. Lanza, D. K. Short, K. R. Short et al., "Endurance exercise as a countermeasure for aging," Diabetes, vol. 57, no. 11, pp. 2933–2942, 2008.

87. J. E. Puche, M. García-Fernández, J. Muntané, J. Rioja, S. González-Barón, and I. C. Cortazar, "Low doses of insulin-like growth factor-I induce mitochondrial protection in aging rats," Endocrinology, vol. 149, no. 5, pp. 2620–2627, 2008.

88. A. Hernández-Aguilera, A. Rull, E. Rodríguez-Gallego et al., "Mitochondrial dysfunction: a basic mechanism in inflammation-related non-communicable diseases and therapeutic opportunities," Mediators of Inflammation, vol. 2013, Article ID 135698, 13 pages, 2013.

89. M. C. Velarde, "Mitochondrial and sex steroid hormone crosstalk during aging," Longevity & Healthspan, vol. 3, no. 1, article 2, 2014.

90. A.-M. G. Psarra and C. E. Sekeris, "Steroid and thyroid hormone receptors in mitochondria," IUBMB Life, vol. 60, no. 4, pp. 210–223, 2008.

91. A.-M. G. Psarra and C. E. Sekeris, "Glucocorticoid receptors and other nuclear transcription factors in mitochondria and possible functions," Biochimica et Biophysica Acta, vol. 1787, no. 5, pp. 431–436, 2009.

92. X. Liu, D. Weaver, O. Shirihai, and G. Hajnóczky, "Mitochondrial 'kiss☐and☐run': interplay between mitochondrial motility and fusion–fission dynamics," The EMBO Journal, vol. 28, no. 20, pp. 3074–3089, 2009.

93. T. Wenz, S. G. Rossi, R. L. Rotundo, B. M. Spiegelman, and C. T. Moraes, "Increased muscle PGC-1α expression protects from sarcopenia and metabolic disease during aging," Proceedings of the National Academy of Sciences of the United States of America, vol. 106, no. 48, pp. 20405–20410, 2009.

94. P. Sestili, E. Barbieri, C. Martinelli et al., "Creatine supplementation prevents the inhibition of myogenic differentiation in oxidatively injured C2C12 murine myoblasts," Molecular Nutrition & Food Research, vol. 53, no. 9, pp. 1187–1204, 2009.

95. E. Barbieri, M. Battistelli, L. Casadei et al., "Morphofunctional and biochemical approaches for studying mitochondrial changes during myoblasts differentiation," Journal of Aging Research, vol. 2011, Article ID 845379, 16 pages, 2011.

96. C. Kang, K. M. O'Moore, J. R. Dickman, and L. L. Ji, "Exercise activation of muscle peroxisome proliferator-activated receptor-gamma coactivator-1alpha signaling is

redox sensitive," Free Radical Biology & Medicine, vol. 47, no. 10, pp. 1394–1400, 2009.

97. I. Irrcher, V. Ljubicic, and D. A. Hood, "Interactions between ROS and AMP kinase activity in the regulation of PGC-1α transcription in skeletal muscle cells," American Journal of Physiology: Cell Physiology, vol. 296, no. 1, pp. C116–C123, 2009.

98. V. Romanello, E. Guadagnin, L. Gomes et al., "Mitochondrial fission and remodelling contributes to muscle atrophy," The EMBO Journal, vol. 29, no. 10, pp. 1774–1785, 2010.

99. M. M. Mihaylova and R. J. Shaw, "The AMPK signalling pathway coordinates cell growth, autophagy and metabolism," Nature Cell Biology, vol. 13, no. 9, pp. 1016–1023, 2011.

100. D. Arnoult, A. Grodet, Y. J. Lee, J. Estaquier, and C. Blackstone, "Release of OPA1 during apoptosis participates in the rapid and complete release of cytochrome c and subsequent mitochondrial fragmentation," The Journal of Biological Chemistry, vol. 280, no. 42, pp. 35742–35750, 2005.

101. A. Misko, S. Jiang, I. Wegorzewska, J. Milbrandt, and R. H. Baloh, "Mitofusin 2 is necessary for transport of axonal mitochondria and interacts with the Miro/Milton complex," The Journal of Neuroscience, vol. 30, no. 12, pp. 4232–4240, 2010.

102. S. Cipolat, T. Rudka, D. Hartmann et al., "Mitochondrial rhomboid PARL regulates cytochrome c release during apoptosis via OPA1-dependent cristae remodeling," Cell, vol. 126, no. 1, pp. 163–175, 2006.

103. Z.-H. Sheng and Q. Cai, "Mitochondrial transport in neurons: impact on synaptic homeostasis and neurodegeneration," Nature Reviews Neuroscience, vol. 13, no. 2, pp. 77–93, 2012.

104. A. M. Cuervo, E. Bergamini, U. T. Brunk, W. Dröge, M. Ffrench, and A. Terman, "Autophagy and aging: the importance of maintaining 'clean' cells," Autophagy, vol. 1, no. 3, pp. 131–140, 2005.

105. E. Masiero and M. Sandri, "Autophagy inhibition induces atrophy and myopathy in adult skeletal muscles," Autophagy, vol. 6, no. 2, pp. 307–309, 2010.

106. T. Vellai, K. Takács-Vellai, M. Sass, and D. J. Klionsky, "The regulation of aging: does autophagy underlie longevity?" Trends in Cell Biology, vol. 19, no. 10, pp. 487–494, 2009.

107. P. Löw, "The role of ubiquitin-proteasome system in ageing," General and Comparative Endocrinology, vol. 172, no. 1, pp. 39–43, 2011.

108. K. G. Lyamzaev, O. K. Nepryakhina, V. B. Saprunova et al., "Novel mechanism of elimination of malfunctioning mitochondria (mitoptosis): formation of mitoptotic bodies and extrusion of mitochondrial material from the cell," Biochimica et Biophysica Acta, vol. 1777, no. 7-8, pp. 817–825, 2008.

109. M. Guescini, D. Guidolin, L. Vallorani et al., "C2C12 myoblasts release micro-vesicles containing mtDNA and proteins involved in signal transduction," Experimental Cell Research, vol. 316, no. 12, pp. 1977–1984, 2010.

110. J. L. Mair, C. A. Boreham, M. Ditroilo et al., "Benefits of a worksite or home-based bench stepping intervention for sedentary middle-aged adults—a pilot study," Clinical Physiology and Functional Imaging, vol. 34, no. 1, pp. 10–17, 2014.

111. K. E. Yarasheski, J. Pak-Loduca, D. L. Hasten, K. A. Obert, M. B. Brown, and D. R. Sinacore, "Resistance exercise training increases mixed muscle protein synthesis

rate in frail women and men ≥76 yr old," The American Journal of Physiology—Endocrinology and Metabolism, vol. 277, no. 1, part 1, pp. E118–E125, 1999.

112. M. A. Fiatarone Singh, W. Ding, T. J. Manfredi et al., "Insulin-like growth factor I in skeletal muscle after weight-lifting exercise in frail elders," The American Journal of Physiology: Endocrinology and Metabolism, vol. 277, no. 1, part 1, pp. E135–E143, 1999.

113. G. Parise, A. N. Brose, and M. A. Tarnopolsky, "Resistance exercise training decreases oxidative damage to DNA and increases cytochrome oxidase activity in older adults," Experimental Gerontology, vol. 40, no. 3, pp. 173–180, 2005.

114. G. F. Martel, S. M. Roth, F. M. Ivey et al., "Age and sex affect human muscle fibre adaptations to heavy-resistance strength training," Experimental Physiology, vol. 91, no. 2, pp. 457–464, 2006.

115. K. Häkkinen, W. J. Kraemer, A. Pakarinen et al., "Effects of heavy resistance/power training on maximal strength, muscle morphology, and hormonal response patterns in 60–75-year-old men and women," Canadian Journal of Applied Physiology, vol. 27, no. 3, pp. 213–231, 2002.

116. E. L. Cadore, R. S. Pinto, M. Bottaro, and M. Izquierdo, "Strength and endurance training prescription in healthy and frail elderly," Aging and Disease, vol. 5, no. 3, pp. 183–195, 2014.

117. T. Ogata and Y. Yamasaki, "Ultra-high-resolution scanning electron microscopy of mitochondria and sarcoplasmic reticulum arrangement in human red, white, and intermediate muscle fibers," The Anatomical Record, vol. 248, no. 2, pp. 214–223, 1997.

118. D. A. Hood, "Invited review: contractile activity-induced mitochondrial biogenesis in skeletal muscle," Journal of Applied Physiology, vol. 90, no. 3, pp. 1137–1157, 2001.

119. E. Barbieri, P. Sestili, L. Vallorani et al., "Mitohormesis in muscle cells: a morphological, molecular, and proteomic approach," Muscles, Ligaments and Tendons Journal, vol. 3, no. 4, pp. 254–266, 2013.

120. Z. Yan, V. A. Lira, and N. P. Greene, "Exercise training-induced regulation of mitochondrial quality," Exercise and Sport Sciences Reviews, vol. 40, no. 3, pp. 159–164, 2012.

121. S. A. Whitman, M. J. Wacker, S. R. Richmond, and M. P. Godard, "Contributions of the ubiquitin-proteasome pathway and apoptosis to human skeletal muscle wasting with age," Pflugers Archiv, vol. 450, no. 6, pp. 437–446, 2005.

122. E. Marzetti, J. C. Y. Hwang, H. A. Lees et al., "Mitochondrial death effectors: relevance to sarcopenia and disuse muscle atrophy," Biochimica et Biophysica Acta, vol. 1800, no. 3, pp. 235–244, 2010.

123. L. L. Ji, "Exercise at old age: does it increase or alleviate oxidative stress?" Annals of the New York Academy of Sciences, vol. 928, pp. 236–247, 2001.

124. N. Eynon, M. Morán, R. Birk, and A. Lucia, "The champions' mitochondria: is it genetically determined? A review on mitochondrial DNA and elite athletic performance," Physiological Genomics, vol. 43, no. 13, pp. 789–798, 2011.

125. C. Bouchard, M. A. Sarzynski, T. K. Rice et al., "Genomic predictors of the maximal O2 uptake response to standardized exercise training programs," Journal of Applied Physiology, vol. 110, no. 5, pp. 1160–1170, 2011.

126. V. Klissouras, "Heritability of adaptive variation," Journal of Applied Physiology, vol. 31, no. 3, pp. 338–344, 1971.

127. D. J. Bishop, C. Granata, and N. Eynon, "Can we optimise the exercise training prescription to maximise improvements in mitochondria function and content?" Biochimica et Biophysica Acta, vol. 1840, no. 4, pp. 1266–1275, 2014.

128. M. A. Tarnopolsky, "Mitochondrial DNA shifting in older adults following resistance exercise training," Applied Physiology, Nutrition and Metabolism, vol. 34, no. 3, pp. 348–354, 2009.

129. S. B. Wilkinson, S. M. Phillips, P. J. Atherton et al., "Differential effects of resistance and endurance exercise in the fed state on signalling molecule phosphorylation and protein synthesis in human muscle," The Journal of Physiology, vol. 586, no. 15, pp. 3701–3717, 2008.

130. J. E. Tang, J. W. Hartman, and S. M. Phillips, "Increased muscle oxidative potential following resistance training induced fibre hypertrophy in young men," Applied Physiology, Nutrition and Metabolism, vol. 31, no. 5, pp. 495–501, 2006.

131. G. Parise, S. M. Phillips, J. J. Kaczor, and M. A. Tarnopolsky, "Antioxidant enzyme activity is up-regulated after unilateral resistance exercise training in older adults," Free Radical Biology & Medicine, vol. 39, no. 2, pp. 289–295, 2005.

132. L. K. Kwong and R. S. Sohal, "Age-related changes in activities of mitochondrial electron transport complexes in various tissues of the mouse," Archives of Biochemistry and Biophysics, vol. 373, no. 1, pp. 16–22, 2000.

133. X. Wang, A. M. Pickrell, S. G. Rossi et al., "Transient systemic mtDNA damage leads to muscle wasting by reducing the satellite cell pool," Human Molecular Genetics, vol. 22, no. 19, pp. 3976–3986, 2013.

134. G. Pallafacchina, S. François, B. Regnault et al., "An adult tissue-specific stem cell in its niche: a gene profiling analysis of in vivo quiescent and activated muscle satellite cells," Stem Cell Research, vol. 4, no. 2, pp. 77–91, 2010.

135. L. Vahidi Ferdousi, P. Rocheteau, R. Chayot et al., "More efficient repair of DNA double-strand breaks in skeletal muscle stem cells compared to their committed progeny," Stem Cell Research A, vol. 13, no. 3, pp. 492–507, 2014.

136. T. Taivassalo, K. Fu, T. Johns, D. Arnold, G. Karpati, and E. A. Shoubridge, "Gene shifting: a novel therapy for mitochondrial myopathy," Human Molecular Genetics, vol. 8, no. 6, pp. 1047–1052, 1999.

137. E. Schultzand and K. M. McCormick, "Skeletal muscle satellite cells," Reviews of Physiology, Biochemistry and Pharmacology, vol. 123, pp. 213–257, 1994.

138. J. L. Murphy, E. L. Blakely, A. M. Schaefer et al., "Resistance training in patients with single, large-scale deletions of mitochondrial DNA," Brain, vol. 131, no. 11, pp. 2832–2840, 2008.

139. S. Spendiff, M. Reza, J. L. Murphy et al., "Mitochondrial DNA deletions in muscle-satellite cells: implications for therapies," Human Molecular Genetics, vol. 22, no. 23, Article ID ddt327, pp. 4739–4747, 2013.

140. L. Wang, H. Mascher, N. Psilander, E. Blomstrand, and K. Sahlin, "Resistance exercise enhances the molecular signaling of mitochondrial biogenesis induced by endurance exercise in human skeletal muscle," Journal of Applied Physiology, vol. 111, no. 5, pp. 1335–1344, 2011.

141. W. Apró, L. Wang, M. Pontén, E. Blomstrand, and K. Sahlin, "Resistance exercise induced mTORC1 signaling is not impaired by subsequent endurance exercise in human skeletal muscle," The American Journal of Physiology: Endocrinology and Metabolism, vol. 305, no. 1, pp. E22–E32, 2013.

142. L. G. MacNeil, E. Glover, T. G. Bergstra, A. Safdar, M. A. Tarnopolsky, and G. L. Lluch, "The order of exercise during concurrent training for rehabilitation does not alter acute genetic expression, mitochondrial enzyme activity or improvements in muscle function," PLoS ONE, vol. 9, no. 10, Article ID e109189, 2014.

143. M. Guescini, C. Fatone, L. Stocchi et al., "Fine needle aspiration coupled with real-time PCR: a painless methodology to study adaptive functional changes in skeletal muscle," Nutrition, Metabolism and Cardiovascular Diseases, vol. 17, no. 5, pp. 383–393, 2007.

144. C. Fatone, M. Guescini, S. Balducci et al., "Two weekly sessions of combined aerobic and resistance exercise are sufficient to provide beneficial effects in subjects with type 2 diabetes mellitus and metabolic syndrome," Journal of Endocrinological Investigation, vol. 33, no. 7, pp. 489–495, 2010.

145. A. A. Turco, M. Guescini, V. Valtucci et al., "Dietary fat differentially modulate the mRNA expression levels of oxidative mitochondrial genes in skeletal muscle of healthy subjects," Nutrition, Metabolism and Cardiovascular Diseases, vol. 24, no. 2, pp. 198–204, 2014.

146. S. Olsen, P. Aagaard, F. Kadi et al., "Creatine supplementation augments the increase in satellite cell and myonuclei number in human skeletal muscle induced by strength training," The Journal of Physiology, vol. 573, part 2, pp. 525–534, 2006.

147. M. C. Devries and S. M. Phillips, "Creatine supplementation during resistance training in older adults—a meta-analysis," Medicine and Science in Sports and Exercise, vol. 46, no. 6, pp. 1194–1203, 2014.

PART III

OTHER THERAPEUTIC STRATEGIES

CHAPTER 9

Novel Intriguing Strategies Attenuating to Sarcopenia

KUNIHIRO SAKUMA AND AKIHIKO YAMAGUCHI

9.1 INTRODUCTION

Skeletal muscle contractions power human body movements and are essential for maintaining stability. Skeletal muscle tissue accounts for almost half of the human body mass and, in addition to its power-generating role, is a crucial factor in maintaining homeostasis. Given its central role in human mobility and metabolic function, any deterioration in the contractile, material, and metabolic properties of skeletal muscle has an extremely important effect on human health. Aging is associated with a progressive decline of muscle mass, quality, and strength, a condition known as sarcopenia [1]. The term sarcopenia, coincd by I. H. Rosenberg, originates from the Greek words sarx (flesh) and penia (loss). Although this term is applied clinically to denote loss of muscle mass, it is often used to describe

Novel Intriguing Strategies Attenuating to Sarcopenia. © *Sakuma K and Yamaguchi A.* Journal of Aging Research *2012 (2012), http://dx.doi.org/10.1155/2012/251217. Licensed under Creative Commons Attribution 3.0 Unported License, http://creativecommons.org/licenses/by/3.0/.*

both a set of cellular processes (denervation, mitochondrial dysfunction, inflammatory and hormonal changes) and a set of outcomes such as decreased muscle strength, decreased mobility and function [2], increased fatigue, a greater risk of falls [3], and reduced energy needs [4]. In addition, reduced muscle mass in aged individuals has been associated with decreased survival rates following critical illness [5]. Estimates of the prevalence of sarcopenia range from 13% to 24% in adults over 60 years of age to more than 50% in persons aged 80 and older [2]. The estimated direct healthcare costs attributable to sarcopenia in the United States in 2000 were $18.5 billion ($10.8 billion in men and $7.7 billion in women), which represented about 1.5% of total healthcare expenditures for that year [6]. Therefore, age-related losses in skeletal muscle mass and function present an extremely important current and future public health issue.

Lean muscle mass generally contributes up to ~50% of total body weight in young adults but declines with aging to be 25% at 75–80 yr old [7, 8]. The loss of muscle mass is typically offset by gains in fat mass. The loss of muscle mass is most notable in the lower limb muscle groups, with the cross-sectional area of the vastus lateralis being reduced by as much as 40% between the age of 20 and 80 yr [9]. On a muscle fiber level, sarcopenia is characterized by specific type II muscle fiber atrophy, fiber necrosis, and fiber-type grouping [9–13]. In elderly men, Verdijk et al. [12] showed a reduction in type II muscle fiber satellite cell content with aging. Although various investigators support such an age-related decrease in the number of satellite cells [12–17], some reports [18–20] indicate no such change. In contrast, most studies point to an age-dependent reduction in muscle-regenerative capacity due to reduced satellite cell proliferation and differentiation.

Several possible mechanisms for age-related muscle atrophy have been described; however, the precise contribution of each is unknown. Age-related muscle loss is a result of reductions in the size and number of muscle fibers [21] possibly due to a multifactorial process that involves physical activity, nutritional intake, oxidative stress, and hormonal changes [3, 22]. The specific contribution of each of these factors is unknown, but there is emerging evidence that the disruption of several positive regulators (Akt and serum response factor) of muscle hypertrophy with age is an important feature in the progression of sarcopenia [23–25]. In contrast, many investigators have failed to demonstrate an age-related enhancement in levels of common negative regulators [atrophy gene-1 (Atrogin-1), myostatin, and calpain] in senescent mammalian muscles [24, 25].

Resistance training combined with amino acid-containing supplements is effective candidate to prevent age-related muscle wasting and weakness [24–26]. In particular, sarcopenia has been most attenuated by treatment with many essential amino acids plus high-amount leucine [24–26]. In addition, many researchers have focused on inhibiting myostatin for treating various muscle disorders such as muscular dystrophy, cachexia, and sarcopenia [27, 28]. Furthermore, more recent studies have indicated a possible application of new supplements to prevent muscle atrophy [29, 30]. This review aims to address several novel strategies for inhibiting the muscle wasting in particular sarcopenia.

9.2 MYOSTATIN INHIBITION

Growth and differentiation factor 8, otherwise known as myostatin, was first discovered during screening for members of a novel transforming growth factor-β (TGF-β) superfamily and shown to act as a potent negative regulator of muscle growth [31, 32]. Studies indicate that myostatin inhibits cell-cycle progression and levels of myogenic regulatory factors, thereby controlling myoblastic proliferation and differentiation during developmental myogenesis [32–35]. Mutations in myostatin can lead to massive hypertrophy and/or hyperplasia in developing animals, as evidenced by knockout experiments in mice and by the phenotype seen in myostatin-null cattle [36] and humans [37]. Myostatin binds to and signals through a combination of Activin IIA/B receptors (ActRIIA/IIB) on the cell membrane; however, it has higher affinity for ActRIIB. On binding ActRIIB, myostatin forms a complex with a second surface type I receptor, either activin receptor-like kinase (ALK4 or ActRIB) or ALK5, to stimulate the phosphorylation of receptor Smad (Rsmad), and Smad2/3 transcription factors in the cytoplasm. Then Smad2/3 translocate and modulate nuclear gene transcription such as MyoD [33] via a TGF-β-like mechanism. In contrast, forkhead box O (FOXO) 1 and Smad2 appear to control the differentiation of C2C12 myoblasts by regulating myostatin mRNA and its promoters [38].

Studies measuring myostatin levels during aging have yielded conflicting results such as marked increases in humans at the mRNA and protein levels [39], no change in mice at the protein level [40], and a decrease in rats at the mRNA level [41]. The functional role of myostatin in aged mammalian muscle may be revealed by further descriptive analysis using other methods (ex. immunofluorescence) and examining the adaptive changes in downstream modulators (ex. ActRIIB, Smad3) of myostatin signaling.

Many researchers have focused on inhibiting myostatin for treating various muscle disorders. The use of neutralizing antibodies to myostatin improved muscle disorders in rodent models of Duchenne muscular dystrophy (DMD), limb girdle muscular dystrophy 2F (Sgcg$^{-/-}$), and amyotrophic lateral sclerosis (SOD1^{G93A} transgenic mouse) [27, 28, 42, 43]. Indeed, myostatin inhibition using MYO-029 was tested in a prospective, randomized, and US phase I/II trial in 116 adults with muscular dystrophy such as Becker muscular dystrophy, facioscapulohumeral muscular dystrophy, and limb-girdle muscular dystrophy [44]. On the other hand, inhibiting myostatin to counteract sarcopenia has also been investigated only in animals. A lack of myostatin caused by gene manipulation increased the number of satellite cells and enlarged the cross-sectional area of predominant type IIB/X fibers in tibialis anterior muscles of mice [45]. In addition, these myostatin-null mice showed prominent regenerative potential including accelerated fiber remodeling after an injection of notexin [45]. LeBrasseur et al. [46] reported several positive effects of 4 weeks of treatment with PF-354 (24 mg/Kg), a drug for myostatin inhibition, in aged mice. They showed that PF-354-treated mice exhibited significantly greater muscle mass (by 12%), and increased performance such as treadmill time, distance to exhaustion, and habitual activity. Furthermore, PF-354-treated mice exhibited decreased levels of phosphorylated Smad3 and muscle ring-finger protein 1 (MuRF1) in aged muscle. More recently, Murphy et al. [47] showed, by way of once weekly injections, that a lower dose of PF-354 (10 mg/Kg) significantly increased the fiber cross-sectional area (by 12%) and in situ force of tibialis anterior muscles (by 35%) of aged mice (21-mo-old). In addition, this form of treatment reduced markers of apoptosis by 56% and reduced caspase3 mRNA levels by 65%. Blocking myostatin enhances muscle protein synthesis [48] by potentially relieving the inhibition normally imposed on the Akt/mammalian target of rapamycin- (mTOR) signaling pathway by myostatin [49]. The blockade may also attenuate muscle protein degradation by inhibiting the ubiquitin-proteasome system, which is controlled, in part, by Akt [50, 51] although the mechanism involved has not been demonstrated. In contrast, a microarray analysis of the skeletal muscle of myostatin knockout mice showed an increased expression of antiapoptotic genes compared with that in control mice [51]. These lines of evidence clearly highlight the therapeutic potential of antibody-directed inhibition of myostatin for treating sarcopenia by inhibiting protein degradation and/or apoptosis.

9.3 URSOLIC ACID

A water-insoluble pentacyclic triterpenoid, ursolic acid is the major waxy component in apple peels [52]. It is also found in many other edible plants. Interestingly, because it exerts beneficial effects in animal models of diabetes and hyperlipidemia [53, 54], ursolic acid is thought to be the active component in a variety of folkloric antidiabetic herbal medicines [53, 55]. As predicted by connectivity mapping, Kunkel et al. [29] found that ursolic acid reduced skeletal muscle atrophy in the setting of two-distinct atrophy-inducing stresses (fasting and muscle denervation). A major strength of the connectivity map is that it takes into account positive and negative changes in mRNA expression that together constitute an authentic mRNA expression signature. Thus, by querying the connective map with signatures of muscle atrophy, Kunkel et al. [29] were, in effect, querying with the reciprocal signature of muscle atrophy but also induced muscle hypertrophy.

Ursolic acid might increase muscle mass by inhibiting atrophy-associated skeletal muscle gene expression. Indeed, Kunkel et al. [29] found that acute ursolic acid treatment of fasted mice reduced Atrogin-1 and MuRF1 mRNA levels in association with reduced muscle atrophy. Similarly, chronic ursolic acid treatment of unstressed mice reduced Atrogin-1 and MuRF1 expression and induced muscle hypertrophy. Although ursolic acid increased skeletal muscle Akt phosphorylation in vivo, the experiments could not determine if it acted directly on skeletal muscle, how quickly it acted, and if the effect required insulin-like growth factor-I (IGF-I) or insulin, which are always present in healthy animals, even during fasting. To address these issues, Kunkel et al. [29] studied serum-starved skeletal myotubes and found that ursolic acid rapidly stimulated IGF-I receptor and insulin receptor activity, but only if IGF-I or insulin was also present. Taken together, their data suggest that ursolic acid first enhances the capacity of preexisting IGF-I and insulin to activate skeletal muscle IGF-I receptors and insulin receptors, respectively. Importantly, ursolic acid alone was not sufficient to increase phosphorylation of the IGF-I receptor or insulin receptor. Rather, its effects also required IGF-I and insulin, respectively. This suggests that ursolic acid either facilitates hormone-mediated receptor autophosphorylation or inhibits receptor dephosphorylation. The latter possibility is supported by previous in vitro data showing that ursolic acid directly inhibits PTP1B [56], a tyrosine phosphatase that dephosphorylates (inactivates) the IGF-I and insulin receptors [57]. Further research is needed to elucidate the effect of supplementation with ursolic acid in skeletal muscle and to attenuate muscle wasting (ex. sarcopenia).

9.4 EICOSAPENTAENOIC ACID

Eicosapentaenoic acid (EPA) is a 20-carbon omega (n)-3 polyunsaturated fatty acid with anti-inflammatory properties, which is synthesized from ingested alpha-linolenic acid or is consumed in fish and fish oil such as cod liver, sardine, and salmon oil [58]. There is no established Dietary Reference Intake for n-3 fatty acids; yet, adequate intake (AI) is set at 1.6 and 1.1 g/d for men and women, respectively. While intake in the United States occurs at levels much lower than the proposed AI and no signs of deficiency are observed, the AI is proposed to provide optimal health benefits associated with consuming omega-3 polyunsaturated fatty acids [59]. Several clinical trials have reported potential health benefits of omega-3 polyunsaturated fatty acids in many diseases, including cardiovascular diseases [60], epilepsy, inflammatory bowel disease, exercise-trained subjects [61], and cancer-associated cachexia [62]. In particular, the administration of omega-3 fatty acids and EPA capsules or supplements with EPA has been shown to be associated with weight stabilization, gains in lean body mass, and improvements in quality of life markers in weight-losing patients with advanced pancreatic cancer. In addition, EPA has also been shown to inhibit the pro-inflammatory transcription factor nuclear factor kappaB (NF-κB) [62, 63], to reduce tumor necrosis factor-α (TNF-α) production by macrophages [64] and to prevent the damaging effects of TNF-α during skeletal muscle differentiation in vitro [65]. Furthermore, short-term treatment with EPA (16 day, 100 mg/Kg) attenuates the muscle degeneration of mdx mice, a model of DMD [66]. EPA treatment decreased creatine kinase levels and attenuated myonecrosis (decrease in Evans-blue dye-positive fibers and a concomitant increase in peripheral nucleated fibers), and reduced the levels of TNF-α.

Some evidence suggests omega-3 polyunsaturated fatty acids to be also a potentially useful therapeutic agent for the treatment and prevention of sarcopenia. In a more recent study [30], sixteen healthy, older adults have been randomly assigned to receive either omega-3 fatty acids or corn oil for 8 week. In their study, the rate of muscle protein synthesis and the phosphorylation of key elements of the anabolic-signaling pathway were evaluated in three different conditions. Smith et al. [30] found that omega-3 fatty acid supplementation had no effect on the basal rate of muscle protein synthesis but augmented the hyperaminoacidemia- hyperinsulinemia-induced increase in the rate of muscle protein synthesis probably due to a greater increase in muscle p70S6K^{Thr389} phosphorylation.

9.5 ANGIOTENSIN-CONVERTING ENZYME INHIBITORS

Angiotensin-converting enzyme (ACE) inhibitors have long been used as a treatment in primary and secondary prevention in cardiovascular disease as well as secondary stroke prevention. It has now been suggested that ACE inhibitors may have a beneficial effect on skeletal muscle. ACE inhibitors may exert their beneficial effects on skeletal muscles through a number of different mechanisms. ACE inhibitors may improve muscle function through improvements in endothelial function, metabolic function, anti-inflammatory effects, and angiogenesis thereby improving skeletal muscle blood flow. ACE inhibitors can increase mitochondrial numbers and IGF-I levels thereby helping to counter sarcopenia [67–69].

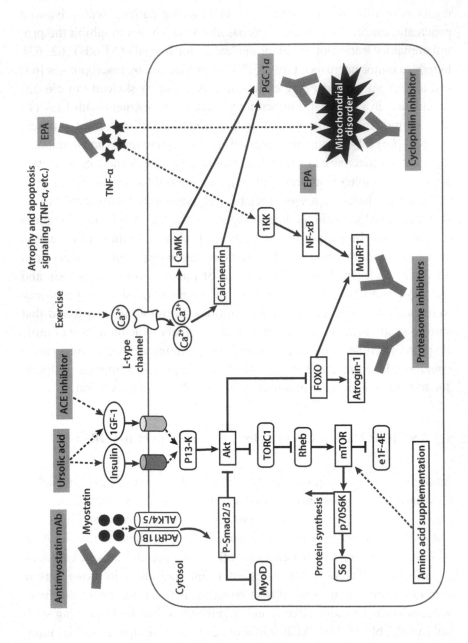

FIGURE 1: Myostatin signals through the ActRIIB-ALK4/5 heterodimer activate Smad2/3 with blocking of MyoD transactivation in an autoregulatory feedback loop. In addition, Smad3 sequesters MyoD in the cytoplasm to prevent it from entering the nucleus and activating the stem cell population. Recent findings [105, 106] have suggested that myostatin-Smad pathway inhibits protein synthesis probably due to blocking the functional role of Akt. Supplementation with ursolic acid upregulates the amount of IGF-I and insulin and then stimulates protein synthesis by activating Akt/mTOR/p70S6K pathway [29]. Treatment with ACE inhibitor also enhances IGF-I level in muscle. Amino acid supplementation enhances protein synthesis by stimulating mTOR [107]. Akt blocks the nuclear translocation of FOXO to inhibit the expression of Atrogin-1 and MuRF1 and the consequent protein degradation. Proteasome inhibitors combat the ubiquitin-proteasome signaling activated by these atrogenes. In cachexic muscle, supplementation with EPA downregulates the amount of TNF-α and NF-κB [63, 64]. Endurance exercise increases the amount of PGC-1α through calcineurin- or CaMK-dependent signaling [108]. Both activated PGC-1α, and cyclophilin inhibitor protects several mitochondrial disorders (apoptosis, oxidative damage, etc.) elicited by the increase in NF-κB and Bax and/or the decrease in Bcl-2 in senescent muscle. ACE; angiotensin-converting enzyme, ActRIIB; activin receptor IIB, ALK4/5; activin-like kinase 4/5, CaMK; calmodulin kinase, eIF4E; eukaryotic initiation factor 4E, EPA; eicosapentaenoic acid, FOXO; Forkhead box O, IGF-I; insulin-like growth factor-I, IKK; inhibitor of κB kinase, mTOR; mammalian target of rapamycin, MuRF1; muscle ring-finger protein 1, NF-κB; nuclear factor of kappa B, PGC-1α; peroxisome proliferator-activated receptor γ coactivator α, PI3-K; phosphatidylinositol 3-kinase, Rheb; Ras homolog enriched in brain, TNF-α; tumor nectosis factor-α, TORC1; a component of TOR-signaling complex 1.

Observational studies have shown that the long-term use of ACE inhibitors was associated with a lower decline in muscle strength and walking speed in older hypertensive people and a greater lower limb lean muscle mass when compared with users of other antihypertensive agents [70]. Several studies have shown that ACE inhibitors improved exercise capacity in both younger and older people with heart failure [70, 71] but caused no improvement in grip strength [72]. Although this could be largely attributed to improvements in cardiac function, skeletal muscle atrophy is also associated with chronic heart failure so the evidence of muscle gains should not be discounted. Few interventional studies using ACE inhibitors for physical function have been undertaken. One study looking at functionally impaired older people without heart failure has shown that ACE inhibitors increase 6-minute walking distance to a degree comparable to that achieved after 6 months of exercise training [73]. Another found that ACE inhibitors increased exercise time in older hypertensive men [74]. However, a study comparing the effects of nifedipine with ACE inhibitors in older people found no difference between treatments in muscle strength, walking distance, or functional performance [75]. It is possible that frailer subjects with slower walking speeds, who have a tendency to more cardiovascular problems, benefit more. Further evidence is required before recommending ACE inhibitors to counter the effects of sarcopenia. However, ACE inhibitors are associated with cardiovascular benefits, and as older people frequently have underlying cardiovascular problems, these agents are already commonly prescribed.

9.6 PROTEASOME INHIBITORS

In a variety of conditions such as cancer, diabetes, denervation, uremia, sepsis, disuse, and fasting, skeletal muscles undergo atrophy through degradation of myofibrillar proteins via the ubiquitin-proteasome pathway [76]. Recent advances have asserted that muscle atrophy in these conditions shares a common mechanism in the induction of the muscle-specific E3 ubiquitin ligases Atrogin-1 and MuRF1 [77–79]. Only very indirect measurements (small increases in mRNA levels encoding some components of the ubiquitin-proteasome pathway [80–82] or ubiquitin-conjugate accumulation [83]) in old muscles of rodents or humans suggested a modest activation of this pathway. Atrogin-1 and/or MuRF1 mRNA levels in aged muscle are reportedly unchanged in humans [83, 84], increased in rats [80, 85], or decreased in rats [82, 86, 87]. When even the mRNA expression of these atrogenes increased in sarcopenic muscles, the induction was very limited (1.5–2.5-fold) as compared with other catabolic situations (10-fold). In addition, the major peptidase activities of the proteasome (i.e., the chymotrypsin-like, trypsin-like, and caspase-like activities) were always reduced (as reported in other tissues [88]) or unchanged with aging [78, 81, 88, 89]. Altogether, these observations clearly suggest that activation of the ubiquitin-proteasome system contributed little to the establishment of sarcopenia in accordance with the very slow muscle mass erosion.

There are several chemical classes of compounds that inhibit proteasomal activity, including peptide analogs of substrates with different C-terminal groups, such as aldehydes, epoxyketones, boronic acids, and vinyl sulfones [90]. A selective boronic acid proteasome inhibitor, Velcade (also known as PS-341 and bortezomib), directly inhibits the proteasome complex without direct effects on ubiquitination. Velcade is well distributed in the body and does not cross the blood-brain barrier. In addition to being useful research tools for dissecting the roles of the proteasome, proteasome inhibitors have potential applications in biotechnology and medicine. For example, through their ability to block the activation of NF-κB, proteasome inhibitors can dramatically reduce in vitro and in vivo the production of inflammatory mediators as well as of various leukocyte adhesion molecules, which play a crucial role in many diseases. Indeed, Velcade is orally active and is presently approved by the Food and Drug Administration and the European Medicines Agency and well tolerated for treating multiple myeloma [91, 92]. Bonuccelli et al. [93] had indicated that Velcade, once injected locally into the gastrocnemius muscles of mdx mice, could upregulate the expression and membrane localization of dystrophin and members of the dystrophin-glycoprotein complex. Gazzerro et al. [94] suggested that treatment with Velcade (0.8 mg/Kg) over a 2-week period reduced muscle degeneration and necrotic features in mdx muscle fibers, as evaluated with Evans blue dye. In addition, they observed many myotubes and/or immature myofibers expressing embryonic myosin heavy chain in mdx muscle after Velcade administration probably due to the upregulation of several myogenic differentiating modulators (MyoD and Myf-5). These effects of Velcade on muscle degeneration would differ dependent on muscle-fiber type. Beehler et al. [95] demonstrated selective attenuation on treatment with Velcade (3 mg/Kg, 7 days) for atrophy of denervated slow-type muscle (soleus), but not fast-type muscle (EDL) of rats. In contrast, MG-132 exerts an inhibitory effect on both the proteasome and the calpain system. More recently, Gazzerro et al. [94] have clearly demonstrated that MG-132 increased dystrophin, alpha-sarcoglycan, and beta-dystroglycan protein levels in explants from Becker muscular dystrophy patients, whereas it increased the proteins of the dystrophin glycoprotein complex in DMD cases. Strangely, there is no rodent study examining the effect of these proteasome inhibitors to prevent muscle atrophy with aging. As indicated previously [80, 82, 88, 89], almost no studies demonstrated an enhancement of proteasome-linked modulators for protein degradation in sarcopenic mammalian muscles. Proteasome inhibitors may not act to attenuate muscle wasting in cases of sarcopenia.

9.7 CYCLOPHILIN INHIBITOR (DEBIO-025)

Ca^{2+} overload is known to cause cellular necrosis by directly inducing the opening of the mitochondrial permeability transition (MPT) pore [96, 97]. The MPT pore spans the inner and outer membranes of the mitochondria and, when opened for prolonged periods of time, leads to loss of ATP generation, swelling, rupture, and induction of cell death [96, 97]. Cyclophilin D is a mitochondrial matrix prolyl cis-trans isomerase that directly regulates calcium—and reactive oxygen species-dependent MPT and cellular necrosis. Indeed, mice lacking Ppif (the gene-encoding cyclophilin D) show protection from necrotic cell death in the brain and heart after ischemic injury, and mitochondria isolated from these mice are resistant to calcium-induced swelling [98, 99]. Additionally, genetic deletion of Ppif attenuated various dystrophic symptoms (fiber atrophy, fiber loss, invasion by inflammatory cells, and swollen mitochondria) of mice lacking δ-sarcoglycan and the α2-chain of laminin-2 [100]. Millay et al. [100] demonstrated that the subcutaneous injection of Debio-025, a potent inhibitor of the cyclophilin family, improved calcium overload-induced swelling of mitochondria and reduced manifestations of necrotic disease such as fibrosis and central nuclei, in mdx mice, a model of DMD. In addition, treatment with Debio-025 prevented mitochondrial dysfunction and normalized the apoptotic rates and ultrastructural lesions of myopathic Co-16a1$^{-/-}$ mice, a model of human Ullrich congenital muscular dystrophy and Bethlem myopathy [101]. More recently, orally administered Debio-025 reduced creatine kinase blood levels and improved grip strength in mdx mice after 6 weeks of treatment [102]. This effect on muscular dystrophy was greater than that of prednisone, currently the standard for treatment of DMD [103, 104]. However, it had not been examined until now whether Debio-025 also has a therapeutic effect on the loss and/or atrophy of muscle fiber with aging in rodents as well as humans. Since there are many symptoms in common between muscular dystrophy and sarcopenia, treatment with Debio-025 may counteract sarcopenic symptoms.

9.8 PGC-1α

Although the mechanisms by which calorie restriction (CR, 30–40%) delays the aging process remain to be fully elucidated, CR is intricately involved in regulating cellular and systemic redox status and in modulating the expression of genes related to macromolecule and organelle turnover, energy metabolism, and cell death and survival [109–111]. Several studies indicate the protection of age-related functional decline and loss of muscle fibers by CR [109, 110]. These protective effects are likely attributable to the ability of CR to reduce the incidence of mitochondrial abnormalities (mitochondrial proton leak), attenuate oxidative stress [reactive oxidative species (ROS) generation], and counteract the age-related increases in proapoptotic signaling in skeletal muscle [109, 112]. Therefore, several lines of evidence suggest mitochondrial involvement in sarcopenia. Therapeutic strategies for sarcopenia like endurance exercise [113] and CR [114] result in increased mitochondrial capacity in the muscle. A key player controlling mitochondrial function is the peroxisome proliferator-activated receptor γ coactivator α (PGC-1α), a master regulator of mitochondrial biogenesis. In skeletal muscle, PGC-1α can also prevent muscle wasting by regulating autophagy [115] and stabilization of the neuromuscular junction program [116] in the context of muscle atrophy during disease. Thereby, PGC-1α links mitochondrial function to muscle integrity [115]. PGC-1α levels in skeletal muscle decrease during aging [116]. The health-promoting effects of increased PGC-1α expression in skeletal muscle have been shown in different mouse models with affected muscle such as DMD [117], denervation-induced atrophy [115], and mitochondrial myopathy [118]. Indeed, Wenz et al. [118] showed that elevated PGC-1α expression in skeletal muscle enhanced oxidative phosphorylation function in a mouse model of mitochondrial myopathy, delaying the onset of the myopathy and markedly prolonging lifespan.

Adipose tissue infiltration of skeletal muscle also increases with age [119, 120]. Sarcopenia may be linked with increased obesity in the elderly [121]. Indeed, persons who are obese and sarcopenic are reported, independent of age, ethnicity, smoking, and comorbidity, to have worse outcomes, including functional impairment, disabilities, and falls, than do those who are nonobese and sarcopenic [122]. Recent work has demonstrated that mitochondrial damage occurs in obese individuals due to enhanced ROS and chronic inflammation caused by increased fatty acid load [123–125]. Specifically, in skeletal muscle, the expression of PGC-1α drives not only mitochondrial biogenesis and the establishment of oxidative myofibers, but also vascularization [126, 127]. It was found that a high-fat diet or fatty acid treatment caused a reduction in the expression of PGC-1α and other mitochondrial genes in skeletal muscle [128], which may be a mechanism through which excess caloric intake impairs skeletal muscle function. A recent study has also demonstrated that transgenic overexpression of PGC-1α in skeletal muscle improved sarcopenia and obesity associated with aging in mice [129]. Endurance training has been shown to upregulate the amount of PGC-1α to elicit mitochondrial biogenesis [130]. The well-known sarcopenia-attenuating effects by endurance training may be attributable to the protection for mitochondrial disorders (apoptosis, oxidative damage, etc.) by the increase of PGC-1α amount. Figure 1 provides an overview of the molecular pathways of muscle hypertrophy and novel strategies for counteracting sarcopenia.

9.9 CONCLUSIONS AND PERSPECTIVES

The advances in our understanding of muscle biology that have occurred over the past decade have led to new hopes for pharmacological treatment of muscle wasting. These treatments will be tested in humans in the coming years and offer the possibility of treating sarcopenia/frailty. These treatments should be developed in the setting of appropriate dietary and exercise strategies. Currently, resistance training combined with amino acid-containing supplements would be the best way to prevent age-related muscle wasting and weakness. Supplementation with ursolic acid and EPA seems to be more intriguing candidates combating sarcopenia although systematic and fundamental research in these treatments has not been conducted even in rodent. The well-known sarcopenia-attenuating effects by endurance training may be attributable to the protection for mitochondrial disorders by the increase of PGC-1α amount.

REFERENCES

1. D. G. Candow and P. D. Chilibeck, "Differences in size, strength, and power of upper and lower body muscle groups in young and older men," Journals of Gerontology Series A: Biological Science Medical Science, vol. 60, no. 2, pp. 148–156, 2005.

2. L. J. Melton 3rd, S. Khosla, C. S. Crowson, W. M. O'Fallon, and B. L. Riggs, "Epidemiology of sarcopenia," Journal of the American Geriatrics Society, vol. 48, no. 6, pp. 625–630, 2000.

3. R. N. Baumgartner, D. L. Waters, D. Gallagher, J. E. Morley, and P. J. Garry, "Predictors of skeletal muscle mass in elderly men and women," Mechanisms of Ageing and Development, vol. 107, no. 2, pp. 123–136, 1999.

4. E. T. Poehlnan, M. J. Toth, and T. Fonong, "Exercise, substrate utilization and energy requirements in the elderly," International Journal of Obesity, vol. 19, no. 4, pp. S93–S96, 1995.

5. R. D. Griffiths, "Muscle mass, survival, and the elderly ICU patient," Nutrition, vol. 12, no. 6, pp. 456–458, 1996.

6. I. Janssen, D. S. Shepard, P. T. Katzmarzyk, and R. Roubenoff, "The healthcare costs of sarcopenia in the United States," Journal of the American Geriatrics Society, vol. 52, no. 1, pp. 80–85, 2004.

7. K. R. Short and K. S. Nair, "The effect of age on protein metabolism," Current Opinion in Clinical Nutrition and Metabolic Care, vol. 3, no. 1, pp. 39–44, 2000.

8. K. R. Short, J. L. Vittone, J. L. Bigelow, D. N. Proctor, and K. S. Nair, "Age and aerobic exercise training effects on whole body and muscle protein metabolism," American Journal of Physiology: Endocrinology and Metabolism, vol. 286, no. 1, pp. E92–E101, 2004.

9. J. Lexell, "Human aging, muscle mass, and fiber type composition," Journals of Gerontology Series A: Biological Science and Medical Science, vol. 50, pp. 11–16, 1995.

10. L. Larsson, "Morphological and functional characteristics of the ageing skeletal muscle in man. A cross-sectional study," Acta Physiologica Scandinavica, Supplement, vol. 457, pp. 1–36, 1978.

11. L. Larsson, B. Sjödin, and J. Karlsson, "Histochemical and biochemical changes in human skeletal muscle with age in sedentary males age 22-65 years," Acta Physiologica Scandinavica, vol. 103, no. 1, pp. 31–39, 1978.

12. L. B. Verdijk, R. Koopman, G. Schaart, K. Meijer, H. H. Savelberg, and L. J. van Loon, "Satellite cell content is specifically reduced in type II skeletal muscle fibers in the elderly," American Journal of Physiology: Endocrinology and Metabolism, vol. 292, no. 1, pp. E151–E157, 2007.

13. L. B. Verdijk, B. G. Gleeson, R. A. M. Jonkers et al., "Skeletal muscle hypertrophy following resistance training is accompanied by a fiber type-specific increase in

satellite cell content in elderly men," Journals of Gerontology Series A: Biological Science and Medical Sciences, vol. 64, no. 3, pp. 332–339, 2009.

14. A. S. Brack, H. Bildsoe, and S. M. Hughes, "Evidence that satellite cell decrement contributes to preferential decline in nuclear number from large fibres during murine age-related muscle atrophy," Journal of Cell Science, vol. 118, no. 20, pp. 4813–4821, 2005.

15. C. A. Collins, P. S. Zammit, A. P. Ruiz, J. E. Morgan, and T. A. Partridge, "A population of myogenic stem cells that survives skeletal muscle aging," Stem Cells, vol. 25, no. 4, pp. 885–894, 2007.

16. K. Day, G. Shefer, A. Shearer, and Z. Yablonka-Reuveni, "The depletion of skeletal muscle satellite cells with age is concomitant with reduced capacity of single progenitors to produce reserve progeny," Developmental Biology, vol. 340, no. 2, pp. 330–343, 2010.

17. G. Shefer, D. P. van de Mark, J. B. Richardson, and Z. Yablonka-Reuveni, "Satellite-cell pool size does matter: defining the myogenic potency of aging skeletal muscle," Developmental Biology, vol. 294, no. 1, pp. 50–66, 2006.

18. I. M. Conboy, M. J. Conboy, G. M. Smythe, and T. A. Rando, "Notch-mediated restoration of regenerative potential to aged muscle," Science, vol. 302, no. 5650, pp. 1575–1577, 2003.

19. S. M. Roth, G. F. Martel, F. M. Ivey et al., "Skeletal muscle satellite cell populations in healthy young and older men and women," Anatomical Record, vol. 260, no. 4, pp. 351–358, 2000.

20. A. J. Wagers and I. M. Conboy, "Cellular and molecular signatures of muscle regeneration: current concepts and controversies in adult myogenesis," Cell, vol. 122, no. 5, pp. 659–667, 2005.

21. J. Lexell, "Ageing and human muscle: observations from Sweden," Canadian Journal of Applied Physiology, vol. 18, no. 1, pp. 2–18, 1993.

22. R. Roubenoff and V. A. Hughes, "Sarcopenia: current concepts," Journals of Gerontology Series A: Biologial Science and Medical Science, vol. 55, no. 12, pp. M716–M724, 2000.

23. K. Sakuma, M. Akiho, H. Nakashima, H. Akima, and M. Yasuhara, "Age-related reductions in expression of serum response factor and myocardin-related transcription factor A in mouse skeletal muscles," Biochimica et Biophysica Acta Molecular Basis of Disease, vol. 1782, no. 7-8, pp. 453–461, 2008.

24. K. Sakuma and A. Yamaguchi, "Molecular mechanisms in aging and current strategies to counteract sarcopenia," Current Aging Science, vol. 3, no. 2, pp. 90–101, 2010.

25. K. Sakuma and A. Yamaguchi, "Sarcopenia: molecular mechanisms and current therapeutic strategy," in Cell Aging, Nova Science Publishers, Huntington, NY, USA, 2011.

26. D. Paddon-Jones and B. B. Rasmussen, "Dietary protein recommendations and the prevention of sarcopenia," Current Opinion in Clinical Nutrition and Metabolic Care, vol. 12, no. 1, pp. 86–90, 2009.

27. L. Bradley, P. J. Yaworsky, and F. S. Walsh, "Myostatin as a therapeutic target for musculoskeletal disease," Cellular and Molecular Life Sciences, vol. 65, no. 14, pp. 2119–2124, 2008.

28. K. Sakuma and A. Yamaguchi, "Inhibitors of myostatin- and proteasome-dependent signaling for attenuating muscle wasting," Recent Patents on Regenerative Medicine, vol. 1, no. 3, pp. 284–298, 2011.

29. S. D. Kunkel, M. Suneja, S. M. Ebert et al., "mRNA expression signatures of human skeletal muscle atrophy identify a natural compound that increases muscle mass," Cell Metabolism, vol. 13, no. 6, pp. 627–638, 2011.

30. G. I. Smith, P. Atherton, D. N. Reeds et al., "Dietary omega-3 fatty acid supplementation increases the rate of muscle protein synthesis in older adults: a randomized controlled trial," American Journal of Clinical Nutrition, vol. 93, no. 2, pp. 402–412, 2011.

31. A. C. McPherron, A. M. Lawler, and S. J. Lee, "Regulation of skeletal muscle mass in mice by a new TGF-β superfamily member," Nature, vol. 387, no. 6628, pp. 83–90, 1997.

32. S. J. Lee, "Regulation of muscle mass by myostatin," Annual Review of Cell and Developmental Biology, vol. 20, pp. 61–86, 2004.

33. B. Langley, M. Thomas, A. Bishop, M. Sharma, S. Gilmour, and R. Kambadur, "Myostatin inhibits myoblast differentiation by down-regulating MyoD expression," Journal of Biological Chemistry, vol. 277, no. 51, pp. 49831–49840, 2002.

34. M. Thomas, B. Langley, C. Berry et al., "Myostatin, a negative regulator of muscle growth, functions by inhibiting myoblast differentiation," Journal of Biological Chemistry, vol. 275, no. 51, pp. 40235–40243, 2000.

35. W. Yang, Y. Zhang, Y. Li, Z. Wu, and D. Zhu, "Myostatin induces cyclin D1 degradation to cause cell cycle arrest through a phosphatidylinositol 3-kinase/AKT/GSK-3β pathway and is antagonized by insulin-like growth factor 1," Journal of Biological Chemistry, vol. 282, no. 6, pp. 3799–3808, 2007.

36. A. C. McPherron and S. J. Lee, "Double muscling in cattle due to mutations in the myostatin gene," Proceedings of the National Academy of Sciences of the United States of America, vol. 94, no. 23, pp. 12457–12461, 1997.

37. M. Schuelke, K. R. Wagner, L. E. Stolz et al., "Myostatin mutation associated with gross muscle hypertrophy in a child," New England Journal of Medicine, vol. 350, no. 26, pp. 2682–2688, 2004.

38. D. L. Allen and T. G. Unterman, "Regulation of myostatin expression and myoblast differentiation by FoxO and SMAD transcription factors," American Journal of Physiology: Cell Physiology, vol. 292, no. 1, pp. C188–C199, 2007.

39. B. Léger, W. Derave, K. de Bock, P. Hespel, and A. P. Russell, "Human sarcopenia reveals an increase in SOCS-3 and myostatin and a reduced efficiency of Akt phosphorylation," Rejuvenation Research, vol. 11, no. 1, pp. 163–175, 2008.

40. M. E. Carlson, M. Hsu, and I. M. Conboy, "Imbalance between pSmad3 and Notch induces CDK inhibitors in old muscle stem cells," Nature, vol. 454, no. 7203, pp. 528–532, 2008.

41. F. Haddad and G. R. Adams, "Aging-sensitive cellular and molecular mechanisms associated with skeletal muscle hypertrophy," Journal of Applied Physiology, vol. 100, no. 4, pp. 1188–1203, 2006.

42. S. Bogdanovich, T. O. Krag, E. R. Barton et al., "Functional improvement of dystrophic muscle by myostatin blockade," Nature, vol. 420, no. 6914, pp. 418–421, 2002.

43. E. L. Holzbaur, D. S. Howland, N. Weber et al., "Myostatin inhibition slows muscle atrophy in rodent models of amyotrophic lateral sclerosis," Neurobiology of Disease, vol. 23, no. 3, pp. 697–707, 2006.

44. K. R. Wagner, J. L. Fleckenstein, A. A. Amato et al., "A phase I/II trial of MYO-029 in adult subjects with muscular dystrophy," Annals of Neurology, vol. 63, no. 5, pp. 561–571, 2008.

45. V. Siriett, L. Platt, M. S. Salerno, N. Ling, R. Kambadur, and M. Sharma, "Prolonged absence of myostatin reduces sarcopenia," Journal of Cellular Physiology, vol. 209, no. 3, pp. 866–873, 2006.

46. N. K. LeBrasseur, T. M. Schelhorn, B. L. Bernardo, P. G. Cosgrove, P. M. Loria, and T. A. Brown, "Myostatin inhibition enhances the effects of exercise on performance and metabolic outcomes in aged mice," Journals of Gerontology Series A: Biological Science and Medical Sciences, vol. 64, no. 9, pp. 940–948, 2009.

47. K. T. Murphy, R. Koopman, T. Naim et al., "Antibody-directed myostatin inhibition in 21-mo-old mice reveals novel roles for myostatin signaling in skeletal muscle structure and function," FASEB Journal, vol. 24, no. 11, pp. 4433–4442, 2010.

48. S. Welle, K. Burgess, and S. Mehta, "Stimulation of skeletal muscle myofibrillar protein synthesis, p70 S6 kinase phosphorylation, and ribosomal protein S6 phosphorylation by inhibition of myostatin in mature mice," American Journal of Physiology: Endocrinology and Metabolism, vol. 296, no. 3, pp. E567–E572, 2009.

49. A. Amirouche, A. C. Durieux, S. Banzet et al., "Down-regulation of Akt/mammalian target of rapamycin signaling pathway in response to myostatin overexpression in skeletal muscle," Endocrinology, vol. 150, no. 1, pp. 286–294, 2009.

50. C. McFarlane, E. Plummer, M. Thomas et al., "Myostatin induces cachexia by activating the ubiquitin proteolytic system through an NF-κB-independent, FoxO1-dependent mechanism," Journal of Cellular Physiology, vol. 209, no. 2, pp. 501–514, 2006.

51. I. Chelh, B. Meunier, B. Picard et al., "Molecular profiles of Quadriceps muscle in myostatin-null mice reveal PI3K and apoptotic pathways as myostatin targets," BMC Genomics, vol. 10, article 196, 13 pages, 2009.

52. R. T. S. Frighetto, R. M. Welendorf, E. N. Nigro, N. Frighetto, and A. C. Siani, "Isolation of ursolic acid from apple peels by high speed counter-current chromatography," Food Chemistry, vol. 106, no. 2, pp. 767–771, 2008.

53. J. Liu, "Pharmacology of oleanolic acid and ursolic acid," Journal of Ethnopharmacology, vol. 49, no. 2, pp. 57–68, 1995.

54. Z. H. Wang, C. C. Hsu, C. N. Huang, and M. C. Yin, "Anti-glycative effects of oleanolic acid and ursolic acid in kidney of diabetic mice," European Journal of Pharmacology, vol. 628, no. 1–3, pp. 255–260, 2010.

55. J. Liu, "Oleanolic acid and ursolic acid: research perspectives," Journal of Ethnopharmacology, vol. 100, no. 1-2, pp. 92–94, 2005.

56. W. Zhang, D. Hong, Y. Zhou et al., "Ursolic acid and its derivative inhibit protein tyrosine phosphatase 1B, enhancing insulin receptor phosphorylation and stimulating glucose uptake," Biochimica et Biophysica Acta, vol. 1760, no. 10, pp. 1505–1512, 2006.

57. K. A. Kenner, E. Anyanwu, J. M. Olefsky, and J. Kusari, "Protein-tyrosine phosphatase 1B is a negative regulator of insulin- and insulin-like growth factor-I-stimulated

signaling," Journal of Biological Chemistry, vol. 271, no. 33, pp. 19810–19816, 1996.

58. L. M. Arterburn, E. B. Hall, and H. Oken, "Distribution, interconversion, and dose response of n-3 fatty acids in humans," American Journal of Clinical Nutrition, vol. 83, no. 6, pp. 1467S–1476S, 2006.

59. K. C. Fearon, M. F. von Meyenfeldt, A. G. Moses et al., "Effect of a protein and energy dense n-3 fatty acid enriched oral supplement on loss of weight and lean tissue in cancer cachexia: a randomised double blind trial," Gut, vol. 52, no. 10, pp. 1479–1486, 2003.

60. C. R. Harper and T. A. Jacobson, "Usefulness of omega-3 fatty acids and the prevention of coronary heart disease," American Journal of Cardiology, vol. 96, no. 11, pp. 1521–1529, 2005.

61. R. J. Bloomer, D. E. Larson, K. H. Fisher-Wellman, A. J. Galpin, and B. K. Schilling, "Effect of eicosapentaenoic and docosahexaenoic acid on resting and exercise-induced inflammatory and oxidative stress biomarkers: a randomized, placebo controlled, cross-over study," Lipids in Health and Disease, vol. 8, article 36, 2009.

62. T. A. Babcock, W. S. Helton, and N. J. Espat, "Eicosapentaenoic acid (EPA): an antiinflammatory ω-3 fat with potential clinical applications," Nutrition, vol. 16, no. 11-12, pp. 1116–1118, 2000.

63. P. Singer, H. Shapiro, M. Theilla, R. Anbar, J. Singer, and J. Cohen, "Anti-inflammatory properties of omega-3 fatty acids in critical illness: novel mechanisms and an integrative perspective," Intensive Care Medicine, vol. 34, no. 9, pp. 1580–1592, 2008.

64. T. A. Babcock, W. S. Helton, D. Hong, and N. J. Espat, "Omega-3 fatty acid lipid emulsion reduces LPS-stimulated macrophage TNF-α production," Surgical Infections, vol. 3, no. 2, pp. 145–149, 2002.

65. P. Magee, S. Pearson, and J. Allen, "The omega-3 fatty acid, eicosapentaenoic acid (EPA), prevents the damaging effects of tumour necrosis factor (TNF)-alpha during murine skeletal muscle cell differentiation," Lipids in Health and Disease, vol. 7, article 24, 2008.

66. R. V. Machado, A. F. Mauricio, A. P. T. Taniguti, R. Ferretti, H. S. Neto, and M. J. Marques, "Eicosapentaenoic acid decreases TNF-α and protects dystrophic muscles of mdx mice from degeneration," Journal of Neuroimmunology, vol. 232, no. 1-2, pp. 145–150, 2011.

67. J. E. Fabre, A. Rivard, M. Magner, M. Silver, and J. M. Isner, "Tissue inhibition of angiotensin-converting enzyme activity stimulates angiogenesis in vivo," Circulation, vol. 99, no. 23, pp. 3043–3049, 1999.

68. E. M. de Cavanagh, B. Piotrkowski, N. Basso et al., "Enalapril and losartan attenuate mitochondrial dysfunction in aged rats," The FASEB Journal, vol. 17, no. 9, pp. 1096–1098, 2003.

69. M. Maggio, G. P. Ceda, F. Lauretani et al., "Relation of angiotensin converting enzyme inhibitor treatment to insulin-like growth factor-1 serum levels in subjects. 65 years of age (the InCHIANTI study)," American Journal of Cardiology, vol. 97, no. 10, pp. 1525–1529, 2006.

70. G. Onder, B. W. J. H. Penninx, R. Balkrishnan et al., "Relation between use of angiotensin-converting enzyme inhibitors and muscle strength and physical function

in older women: an observational study," Lancet, vol. 359, no. 9310, pp. 926–930, 2002.

71. L. Dossegger, E. Aldor, M. G. Baird et al., "Influence of angiotensin converting enzyme inhibition on exercise performance and clinical symptoms in chronic heart failure—a multicentre, double-blind, placebo-controlled trial," European Heart Journal, vol. 14, pp. 18–23, 1993.

72. G. D. Schellenbaum, N. L. Smith, S. R. Heckbert et al., "Weight loss, muscle strength, and angiotensin-converting enzyme inhibitors in older adults with congestive heart failure or hypertension," Journal of the American Geriatrics Society, vol. 53, no. 11, pp. 1996–2000, 2005.

73. D. Sumukadas, M. D. Witham, A. D. Struthers, and M. E. T. McMurdo, "Effect of perindopril on physical function in elderly people with functional impairment: a randomized controlled trial," Canadian Medical Association Journal, vol. 177, no. 8, pp. 867–874, 2007.

74. G. Leonetti, C. Mazzola, C. Pasotti et al., "Treatment of hypertension in the elderly—effects on blood pressure, heart rate, and physical fitness," American Journal of Medicine, vol. 90, pp. S12–S13, 1991.

75. D. Bunout, G. Barrera, P. M. de la Maza, L. Leiva, C. Backhouse, and S. Hirsch, "Effects of enalapril or nifedipine on muscle strength or functional capacity in elderly subjects. A double blind trial," Journal of the Renin-Angiotensin-Aldosterone System, vol. 10, no. 2, pp. 77–84, 2009.

76. D. Cai, J. D. Frantz, N. E. Tawa Jr. et al., "IKKβ/NF-κB activation causes severe muscle wasting in mice," Cell, vol. 119, no. 2, pp. 285–298, 2004.

77. W. E. Mitch and A. L. Goldberg, "Mechanisms of disease: mechanisms of muscle wasting: the role of the ubiquitin-proteasome pathway," New England Journal of Medicine, vol. 335, no. 25, pp. 1897–1905, 1996.

78. M. Sandri, C. Sandri, A. Gilbert et al., "Foxo transcription factors induce the atrophy-related ubiquitin ligase atrogin-1 and cause skeletal muscle atrophy," Cell, vol. 117, no. 3, pp. 399–412, 2004.

79. T. N. Stitt, D. Drujan, B. A. Clarke et al., "The IGF-1/PI3K/Akt pathway prevents expression of muscle atrophy-induced ubiquitin ligases by inhibiting FOXO transcription factors," Molecular Cell, vol. 14, no. 3, pp. 395–403, 2004.

80. M. Bossola, F. Pacelli, P. Costelli, A. Tortorelli, F. Rosa, and G. B. Doglietto, "Proteasome activities in the rectus abdominis muscle of young and older individuals," Biogerontology, vol. 9, no. 4, pp. 261–268, 2008.

81. J. S. Pattison, L. C. Folk, R. W. Madsen, T. E. Childs, and F. W. Booth, "Transcriptional profiling identifies extensive downregulation of extracellular matrix gene expression in sarcopenic rat soleus muscle," Physiological Genomics, vol. 15, pp. 34–43, 2004.

82. L. Combaret, D. Dardevet, D. Béchet, D. Taillandier, L. Mosoni, and D. Attaix, "Skeletal muscle proteolysis in aging," Current Opinion in Clinical Nutrition and Metabolic Care, vol. 12, no. 1, pp. 37–41, 2009.

83. K. C. DeRuisseau, A. N. Kavazis, and S. K. Powers, "Selective downregulation of ubiquitin conjugation cascade mRNA occurs in the senescent rat soleus muscle," Experimental Gerontology, vol. 40, no. 6, pp. 526–531, 2005.

84. S. Welle, A. L. Brooks, J. M. Delehanty, N. Needler, and C. A. Thornton, "Gene expression profile of aging in human muscle," Physiological Genomics, vol. 14, pp. 149–159, 2003.

85. S. A. Whitman, M. J. Wacker, S. R. Richmond, and M. P. Godard, "Contributions of the ubiquitin-proteasome pathway and apoptosis to human skeletal muscle wasting with age," Pflügers Archiv European Journal of Physiology, vol. 450, no. 6, pp. 437–446, 2005.

86. S. Clavel, A. S. Coldefy, E. Kurkdjian, J. Salles, I. Margaritis, and B. Derijard, "Atrophy-related ubiquitin ligases, atrogin-1 and MuRF1 are up-regulated in aged rat Tibialis Anterior muscle," Mechanisms of Ageing and Development, vol. 127, no. 10, pp. 794–801, 2006.

87. E. Edström, M. Altun, M. Hägglund, and B. Ulfhake, "Atrogin-1/MAFbx and MuRF1 are downregulated in aging-related loss of skeletal muscle," Journals of Gerontology: Series A Biological Sciences and Medical Sciences, vol. 61, no. 7, pp. 663–674, 2006.

88. D. Attaix, L. Mosoni, D. Dardevet, L. Combaret, P. P. Mirand, and J. Grizard, "Altered responses in skeletal muscle protein turnover during aging in anabolic and catabolic periods," International Journal of Biochemistry and Cell Biology, vol. 37, no. 10, pp. 1962–1973, 2005.

89. A. D. Husom, E. A. Peters, E. A. Kolling, N. A. Fugere, L. V. Thompson, and D. A. Ferrington, "Altered proteasome function and subunit composition in aged muscle," Archives of Biochemistry and Biophysics, vol. 421, no. 1, pp. 67–76, 2004.

90. D. H. Lee, "Proteasome inhibitors: valuable new tools for cell biologists," Trends in Cell Biology, vol. 8, no. 10, pp. 397–403, 1998.

91. J. Adams, V. J. Palombella, E. A. Sausville et al., "Proteasome inhibitors: a novel class of potent and effective antitumor agents," Cancer Research, vol. 59, no. 11, pp. 2615–2622, 1999.

92. R. Z. Orlowski, "Proteasome inhibitors in cancer therapy," Methods in Molecular Biology, vol. 100, pp. 197–203, 1997.

93. G. Bonuccelli, F. Sotgia, E. Capozza, E. Gazzerro, C. Minetti, and M. P. Lisanti, "Localized treatment with a novel FDA-approved proteasome inhibitor blocks the degradation of dystrophin and dystrophin-associated proteins in mdx mice," Cell Cycle, vol. 6, no. 10, pp. 1242–1248, 2007.

94. E. Gazzerro, S. Assereto, A. Bonetto et al., "Therapeutic potential of proteasome inhibition in Duchenne and Becker muscular dystrophies," American Journal of Pathology, vol. 176, no. 4, pp. 1863–1877, 2010.

95. B. C. Beehler, P. G. Sleph, L. Benmassaoud, and G. J. Grover, "Reduction of skeletal muscle atrophy by a proteasome inhibitor in a rat model of denervation," Experimental Biology and Medicine, vol. 231, no. 3, pp. 335–341, 2006.

96. P. Bernardi, "Mitochondrial transport of cations: channels, exchangers, and permeability transition," Physiological Reviews, vol. 79, no. 4, pp. 1127–1155, 1999.

97. N. Zamzami and G. Kroemer, "The mitochondrion in apoptosis: how Pandora's box opens," Nature Reviews Molecular Cell Biology, vol. 2, no. 1, pp. 67–71, 2001.

98. T. Nakagawa, S. Shimizu, T. Watanabe et al., "Cyclophilin D-dependent mitochondrial permeability transition regulates some necrotic but not apoptotic cell death," Nature, vol. 434, no. 7033, pp. 652–658, 2005.

99. A. C. Schinzel, O. Takeuchi, Z. Huang et al., "Cyclophilin D is a component of mitochondrial permeability transition and mediates neuronal cell death after focal cerebral ischemia," Proceedings of the National Academy of Sciences of the United States of America, vol. 102, no. 34, pp. 12005–12010, 2005.

100. D. P. Millay, M. A. Sargent, H. Osinska et al., "Genetic and pharmacologic inhibition of mitochondrial-dependent necrosis attenuates muscular dystrophy," Nature Medicine, vol. 14, no. 4, pp. 442–447, 2008.

101. E. R. Wissing, D. P. Millay, G. Vuagniaux, and J. D. Molkentin, "Debio-025 is more effective than prednisone in reducing muscular pathology in mdx mice," Neuromuscular Disorders, vol. 20, no. 11, pp. 753–760, 2010.

102. T. Tiepolo, A. Angelin, E. Palma et al., "The cyclophilin inhibitor Debio 025 normalizes mitochondrial function, muscle apoptosis and ultrastructural defects in Col6a1-/-myopathic mice," British Journal of Pharmacology, vol. 157, no. 6, pp. 1045–1052, 2009.

103. C. Angelini, "The role of corticosteroids in muscular dystrophy: a critical appraisal," Muscle and Nerve, vol. 36, no. 4, pp. 424–435, 2007.

104. B. Balaban, D. J. Matthews, G. H. Clayton, and T. Carry, "Corticosteroid treatment and functional improvement in Duchenne muscular dystrophy long-term effect," American Journal of Physical Medicine and Rehabilitation, vol. 84, no. 11, pp. 843–850, 2005.

105. M. R. Morissette, S. A. Cook, C. Buranasombati, M. A. Rosenberg, and A. Rosenzweig, "Myostatin inhibits IGF-I-induced myotube hypertrophy through Akt," American Journal of Physiology: Cell Physiology, vol. 297, no. 5, pp. C1124–C1132, 2009.

106. A. U. Trendelenburg, A. Meyer, D. Rohner, J. Boyle, S. Hatakeyama, and D. J. Glass, "Myostatin reduces Akt/TORC1/p70S6K signaling, inhibiting myoblast differentiation and myotube size," American Journal of Physiology: Cell Physiology, vol. 296, no. 6, pp. C1258–C1270, 2009.

107. K. E. Wellen and C. B. Thompson, "Cellular metabolic stress: considering how cells respond to nutrient excess," Molecular Cell, vol. 40, no. 2, pp. 323–332, 2010.

108. J. O. Holloszy, "Regulation by exercise of skeletal muscle," Journal of Physiology and Pharmacology, vol. 59, no. 7, pp. 5–18, 2008.

109. A. J. Dirks and C. Leeuwenburgh, "Aging and lifelong calorie restriction result in adaptations of skeletal muscle apoptosis repressor, apoptosis-inducing factor, X-linked inhibitor of apoptosis, caspase-3, and caspase-12," Free Radical Biology and Medicine, vol. 36, no. 1, pp. 27–39, 2004.

110. G. López-Lluch, N. Hunt, B. Jones et al., "Calorie restriction induces mitochondrial biogenesis and bioenergetic efficiency," Proceedings of the National Academy of Sciences of the United States of America, vol. 103, no. 6, pp. 1768–1773, 2006.

111. A. M. Payne, S. L. Dodd, and C. Leeuwenburgh, "Life-long calorie restriction in Fischer 344 rats attenuates age-related loss in skeletal muscle-specific force and reduces extracellular space," Journal of Applied Physiology, vol. 95, no. 6, pp. 2554–2562, 2003.

112. T. Philips and C. Leeuwenburgh, "Muscle fiber specific apoptosis and TNF-α signaling in sarcopenia are attenuated by life-long calorie restriction," FASEB Journal, vol. 19, no. 6, pp. 668–670, 2005.

113. D. R. Taaffe, "Sarcopenia—exercise as a treatment strategy," Australian Family Physician, vol. 35, no. 3, pp. 130–134, 2006.

114. E. Marzetti, H. A. Lees, S. E. Wohlgemuth, and C. Leeuwenburgh, "Sarcopenia of aging: underlying cellular mechanisms and protection by calorie restriction," Bio-Factors, vol. 35, no. 1, pp. 28–35, 2009.

115. M. Sandri, J. Lin, C. Handschin et al., "PGC-1α protects skeletal muscle from atrophy by suppressing FoxO3 action and atrophy-specific gene transcription," Proceedings of the National Academy of Sciences of the United States of America, vol. 103, no. 44, pp. 16260–16265, 2006.

116. R. Anderson and T. Prolla, "PGC-1α in aging and anti-aging interventions," Biochimica et Biophysica Acta, vol. 1790, no. 10, pp. 1059–1066, 2009.

117. C. Handschin, Y. M. Kobayashi, S. Chin, P. Seale, K. P. Campbell, and B. M. Spiegelman, "PGC-1α regulates the neuromuscular junction program and ameliorates Duchenne muscular dystrophy," Genes and Development, vol. 21, no. 7, pp. 770–783, 2007.

118. T. Wenz, F. Diaz, D. Hernandez, et al., "Activation of the PPAR/PGC-1α pathway prevents a bioenergetic deficit and effectively improves a mitochondrial myopathy," Journal of Applied Physiology, vol. 106, pp. 1712–1719, 2008.

119. B. H. Goodpaster, C. L. Carlson, M. Visser et al., "Attenuation of skeletal muscle and strength in the elderly: the health ABC study," Journal of Applied Physiology, vol. 90, no. 6, pp. 2157–2165, 2001.

120. M.-Y. Song, E. Ruts, J. Kim, I. Janumala, S. Heymsfield, and D. Gallagher, "Sarcopenia and increased adipose tissue infiltration of muscle in elderly African American women," American Journal of Clinical Nutrition, vol. 79, no. 5, pp. 874–880, 2004.

121. M. Zamboni, G. Mazzali, F. Fantin, A. Rossi, and V. di Francesco, "Sarcopenic obesity: a new category of obesity in the elderly," Nutrition, Metabolism and Cardiovascular Diseases, vol. 18, no. 5, pp. 388–395, 2008.

122. J. E. Morley, R. N. Baumgartner, R. Roubenoff, J. Mayer, and K. S. Nair, "Sarcopenia," Journal of Laboratory and Clinical Medicine, vol. 137, no. 4, pp. 231–243, 2001.

123. R. J. Roth, A. M. Le, L. Zhang et al., "MAPK phosphatase-1 facilitates the loss of oxidative myofibers associated with obesity in mice," Journal of Clinical Investigation, vol. 119, no. 12, pp. 3817–3829, 2009.

124. E. J. Anderson, M. E. Lustig, K. E. Boyle et al., "Mitochondrial H2O2 emission and cellular redox state link excess fat intake to insulin resistance in both rodents and humans," Journal of Clinical Investigation, vol. 119, no. 3, pp. 573–581, 2009.

125. C. Bonnard, A. Durand, S. Peyrol et al., "Mitochondrial dysfunction results from oxidative stress in the skeletal muscle of diet-induced insulin-resistant mice," Journal of Clinical Investigation, vol. 118, no. 2, pp. 789–800, 2008.

126. Z. Arany, S. Y. Foo, Y. Ma et al., "HIF-independent regulation of VEGF and angiogenesis by the transcriptional coactivator PGC-1α," Nature, vol. 451, no. 7181, pp. 1008–1012, 2008.

127. C. Handschin, S. Chin, P. Li et al., "Skeletal muscle fiber-type switching, exercise intolerance, and myopathy in PGC-1α muscle-specific knock-out animals," Journal of Biological Chemistry, vol. 282, no. 41, pp. 30014–30021, 2007.

128. S. Crunkhorn, F. Dearie, C. Mantzoros et al., "Peroxisome proliferator activator receptor γ coactivator-1 expression is reduced in obesity: potential pathogenic role of saturated fatty acids and p38 mitogen-activated protein kinase activation," Journal of Biological Chemistry, vol. 282, no. 21, pp. 15439–15450, 2007.
129. T. Wenz, S. G. Rossi, R. L. Rotundo, B. M. Spiegelman, and C. T. Moraes, "Increased muscle PGC-1α expression protects from sarcopenia and metabolic disease during aging," Proceedings of the National Academy of Sciences of the United States of America, vol. 106, no. 48, pp. 20405–20410, 2009.
130. Z. Arany, "PGC-1 coactivators and skeletal muscle adaptations in health and disease," Current Opinion in Genetics and Development, vol. 18, no. 5, pp. 426–434, 2008.

CHAPTER 10

Muscle Wasting and Resistance of Muscle Anabolism: The "Anabolic Threshold Concept" for Adapted Nutritional Strategies during Sarcopenia

DOMINIQUE DARDEVET, DIDIER RÉMOND,
MARIE-AGNÈS PEYRON, ISABELLE PAPET,
ISABELLE SAVARY-AUZELOUX, AND LAURENT MOSONI

The main function of skeletal muscle is to provide power and strength for locomotion and posture, but this tissue is also the major reservoir of body proteins and amino acids. Thus, although the loss of muscle proteins has positive effects in the short term by providing amino acids to other tissues, an uncontrolled and sustained muscle wasting impairs movement, leads to difficulties in performing daily activities, and has detrimental metabolic consequences with reduced ability in mobilizing enough amino acids in case of illness and diseases. The resulting weakness increases the incidence of falls and the length of recovery and when advanced, muscle wast-

Muscle Wasting and Resistance of Muscle Anabolism: The "Anabolic Threshold Concept" for Adapted Nutritional Strategies during Sarcopenia. © *Dardevet D, Rémond D, Peyron M-A, Papet I, Savary-Auzeloux I, and Mosoni L.* The Scientific World Journal *2012 (2012). http://dx.doi. org/10.1100/2012/269531. Licensed under a Creative Commons Attribution 3.0 Unported License, http://creativecommons.org/licenses/by/3.0/.*

ing is correlated to morbidity and increased mortality. Consequently, one of the challenges we have to face is to supply amino acids to the tissues with higher requirements in catabolic states [1] but also to prevent a too important loss in muscle proteins and ultimately improve muscle recovery.

During the day, protein metabolism is modified by food intake. Whole-body proteins are stored during postprandial periods and lost in postabsorptive periods. With a muscle protein mass that remains constant, the loss of muscle proteins is compensated for the same protein gain in the postprandial state. In adult volunteers, oral feeding is associated with an increase in whole-body protein synthesis and a decrease in proteolysis [2–5]. These changes are mediated by feeding-induced increases in plasma concentrations of both nutrients and hormones. Many studies suggest that amino acids and insulin play major roles in promoting postprandial protein anabolism [6]. Thus, in case of muscle wasting, muscle protein loss results from an imbalance between protein accretion and break-down rates which, in part, comes from a defect in the postprandial anabolism.

Although each muscle wasting situation is characterized by its specific mechanism(s) and pathways leading to muscle loss, an increase of catabolic factors such as glucocorticoids, cytokines, and oxidative stress, often occurs and it is now well established that these factors have potential deleterious effects on the amino acids or insulin signalling pathways involved in the stimulation of muscle anabolism after food intake [7–11].

These signalling alterations lead to an "anabolic resistance" of muscle even if the anabolic factor requirements (amino acids e.g.) are theoretically covered, that is, with a normal nutrient availability fitting the recommended dietary protein allowances in healthy subjects. This anabolic resistance may be in part explained by an increase of the muscle "anabolic threshold" required to promote maximal anabolism and protein retention (Figures 1(a) and 1(b)). Because the muscle "anabolic threshold" is higher, the anabolic stimuli (including aminoacidemia) cannot reach the anabolic threshold anymore and by consequence, muscle anabolism is reduced with the usual nutrient intake (Figure 1(b)). A possible nutritional strategy is then to increase the intake of anabolic factors (especially amino acids) to reach the new "anabolic threshold" (Figure 1(c)). There are several ways to increase amino acid availability to skeletal muscle: increase protein intake, to supplement the diet with one or several free amino acids or to select the protein source on its amino acid composition and physicochemical properties when digested in the digestive tract. These nutritional strategies tested to increase postprandial amino acid levels above the increased anabolic threshold and ultimately to restimulate muscle protein synthesis in situations of anabolic resistance led to conflicting results with no or more or less positive effects of the supplementation on nitrogen retention. This could be explained by variations in amino acid kinetics. The duration of the hyperaminoacidemia postprandially can also be of a variable magnitude and duration, depending of the form of the protein/amino acid supply in the diet. To illustrate this concept, we will take one example, that is, the loss of muscle mass during aging, while keeping in mind that this could be translated to any situation of muscle wasting.

Sarcopenia, as other catabolic states, has been found to result from a decreased response and/or sensitivity of protein synthesis and degradation to physiologic concentrations of amino acids [12–14]. This is related to a defect of the leucine signal to stimulate the mTOR signalling pathway activity [15]. These data suggest that increasing leucine availability may then represent a nutritional strategy to overcome the "anabolic threshold" increase observed during aging. Studies in both elderly humans and rodents subjected to free leucine supplementation have shown that such supplementations indeed acutely improved muscle protein balance after food intake by increasing muscle protein synthesis and decreasing muscle proteolysis in the postprandial state (reviewed in Balage and Dardevet [16]). However, the few chronic studies conducted with such free leucine supplementations did not succeed in promoting an increase in muscle mass [17–19]. Choosing free leucine as a supplement over a normal protein diet creates a desynchronization between leucine signal and the rise in all amino acids (Figure 2(a)). Indeed the free leucine is absorbed immediately whereas the other amino acids are released later after gastric emptying and proteolytic digestion in the gut. This nonsynchronization between the stimulation of muscle leucine-associated protein metabolism pathways and the delayed availability of amino acids as substrates can explain that protein anabolism was only stimulated for a very short period of time during the postprandial period and then could not translate into a significant muscle protein accretion.

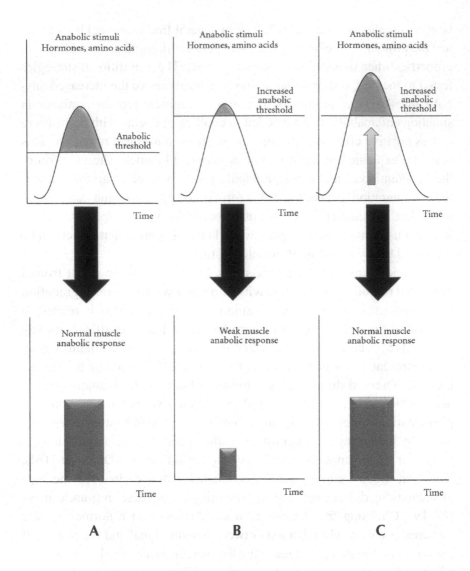

FIGURE 1: The concept of increased anabolic threshold with associated altered muscle protein anabolism during the postprandial period.

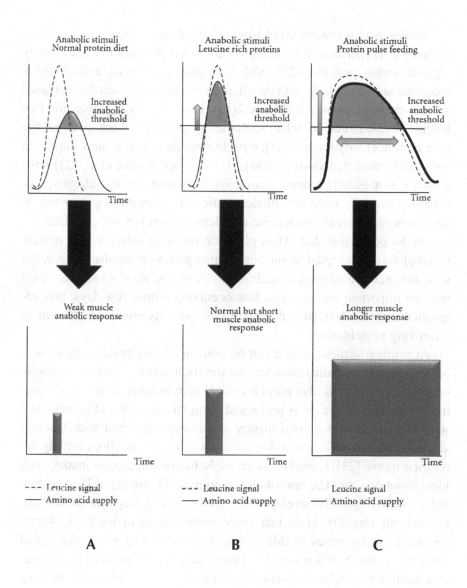

FIGURE 2: Free leucine, leucine rich proteins, and high protein diet in terms of amino acid kinetic and associated anabolic response in situation of increased muscle anabolic threshold.

Studies with a synchronized leucine signal and amino acid availability have been performed by using leucine rich proteins that are rapidly digested (whey proteins) [20]. With such proteins, leucine availability is increased simultaneously with the other amino acids to reach the increased muscle anabolic threshold (Figure 2(b)). However, as observed for free leucine supplementation, when such dietary proteins where given on the long term in elderly rodents [21], muscle anabolism was acutely improved but muscle mass remained unchanged. However, Magne et al. [22] have shown that in elderly rodents recovering from acute muscle atrophy, leucine rich proteins were nevertheless efficient in improving recovery of muscle mass whereas free leucine supplementation remained ineffective. It may be postulated that, when given on the long term, protein muscle metabolism was adapted by increasing also protein catabolism in parallel with the increase of protein anabolism. However, after a catabolic state with an important muscle mass loss occurring within few days, this adaptation may be delayed and leucine rich proteins remained efficient in improving muscle mass.

According to these data, it can be concluded that besides counteracting the muscle anabolic resistance, the duration during which the anabolic resistance is muzzled also plays a critical role in leading to a significant muscle protein accretion. A prolonged stimulation could not be achieved with fast proteins at normal dietary level (even enriched with leucine) since the concentration of amino acids as substrates declines rapidly after their intake [23]. However, by strongly increasing protein intake, such ideal situations could be nevertheless achieved (Figure 2(c)). The "protein pulse feeding" initially developed by Arnal et al. [24–26] have shown that, by concentrating 80% of the total daily protein intake in one meal, protein retention was improved in elderly women subjected to a such nutritional strategy. Similarly, when very large amount of amino acids (wherein leucine formed the highest percentage of the mixture), positive results have been observed [27–30].

The above nutritional strategies discussed raised the problem that the organism has to cope with large amount of nitrogen to eliminate. This point can be critical with already frail sarcopenic subjects or patients for who the renal function will be oversolicited whereas it may be already altered.

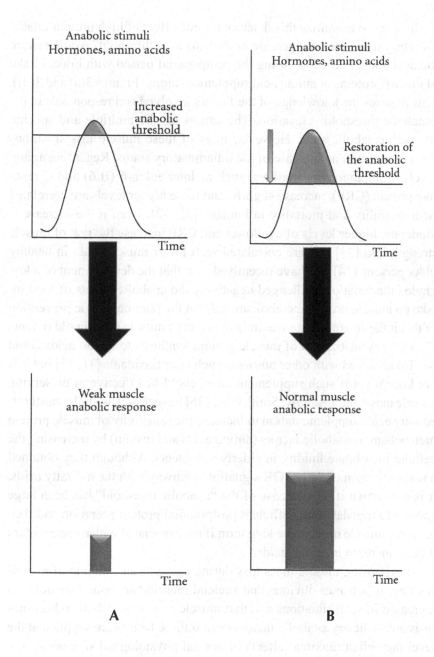

FIGURE 3: Strategies aiming at partially decreasing the muscle "anabolic threshold" and increasing the efficiency of the postprandial period.

In order to minimize this deleterious side effect of high protein intake, a strategy to reverse the increase in the "anabolic threshold" would restore the anabolic stimulation during the postprandial period with lower intake of dietary proteins or amino acid supplementations (Figures 3(a) and 3(b)). This requires the knowledge of the factors involved and responsible in the "anabolic threshold" elevation. The causes can be multiple and specific for each catabolic state. However, most of these muscle loss situations have in common an increase of the inflammatory status. Regarding aging, levels of inflammatory markers, such as interleukin-6 (IL6) and C reactive protein (CRP), increase slightly, and these higher levels are correlated with disability and mortality in humans [31, 32]. Even if the increase is moderate, higher levels of cytokines and CRP increase the risk of muscle strength loss [33] and are correlated with lower muscle mass in healthy older persons [34]. We have recently shown that the development of a low grade inflammation challenged negatively the anabolic effects of food intake on muscle protein metabolism and that the pharmacologic prevention of this inflammatory state was able to preserve muscle mass in old rodents [7, 8]. A resensitization of muscle protein synthesis to amino acids could be also achieved with other nutrients such as antioxidants [11, 35] but it is not known yet if such supplementations could be effective in preserving muscle mass. Interestingly, Smith et al. [36] have tested n-3 polyunsaturated fatty acids supplementation to increase the sensitivity of muscle protein metabolism to anabolic factors (amino acids and insulin) by increasing the cellular membrane fluidity in elderly volunteers. Although they obtained a resensitization of the mTOR signalling pathway with the n-3 fatty acids, it is not known if the decrease of the "anabolic threshold" has been large enough to translate into sufficient postprandial protein accretion and then preserve muscle mass in the long term if not associated with a concomitant dietary increase in amino acids.

By choosing muscle mass loss during aging as an example of muscle wasting, it becomes obvious that skeletal muscle "anabolic threshold" is increased in such situations and that muscle protein metabolism becomes resistant to dietary anabolic factors even if these factors are supplied at the level they elicit maximal effects in normal physiological situations. It is important to note that this anabolic resistance during aging may be specific to amino acids [37]. Because the muscle "anabolic threshold" is more elevated, the duration of the stimulation by anabolic signals (as leucine) and the overcome of amino acid supply above the threshold is reduced with usual nutrient intake. Two strategies can be used (alone or in combination) to deal with this decreased "efficient" postprandial period: (1) by increasing the anabolic signals, and particular amino acid availability; however, it is necessary to synchronize the anabolic stimuli with the substrates in order to optimize the incorporation of amino acids into muscle proteins; (2) by increasing the efficiency of the postprandial period with strategies aiming at partially restoring (i.e., decreasing) the muscle "anabolic threshold".

REFERENCES

1. C. Obled, I. Papet, and D. Breuillé, "Metabolic bases of amino acid requirements in acute diseases," Current Opinion in Clinical Nutrition & Metabolic Care, vol. 5, pp. 189–197, 2002.
2. M. J. Rennie, R. H. T. Edwards, and D. Halliday, "Muscle protein synthesis measured by stable isotope techniques in man: the effects of feeding and fasting," Clinical Science, vol. 63, no. 6, pp. 519–523, 1982.
3. P. J. Pacy, G. M. Price, D. Halliday, M. R. Quevedo, and D. J. Millward, "Nitrogen homoeostasis in man: the diurnal responses of protein synthesis and degradation and amino acid oxidation to diets with increasing protein intakes," Clinical Science, vol. 86, no. 1, pp. 103–118, 1994.
4. Y. Boirie, P. Gachon, S. Corny, J. Fauquant, J. L. Maubois, and B. Beaufrère, "Acute postprandial changes in leucine metabolism as assessed with an intrinsically labeled milk protein," American Journal of Physiology, vol. 271, no. 6, pp. E1083–E1091, 1996.
5. E. Volpi, P. Lucidi, G. Cruciani et al., "Contribution of amino acids and insulin to protein anabolism during meal absorption," Diabetes, vol. 45, no. 9, pp. 1245–1252, 1996.
6. M. Prod'homme, I. Rieu, M. Balage, D. Dardevet, and J. Grizard, "Insulin and amino acids both strongly participate to the regulation of protein metabolism," Current Opinion in Clinical Nutrition and Metabolic Care, vol. 7, no. 1, pp. 71–77, 2004.
7. M. Balage, J. Averous, D. Rémond et al., "Presence of low-grade inflammation impaired postprandial stimulation of muscle protein synthesis in old rats," Journal of Nutritional Biochemistry, vol. 21, no. 4, pp. 325–331, 2010.
8. I. Rieu, C. Sornet, J. Grizard, and D. Dardevet, "Glucocorticoid excess induces a prolonged leucine resistance on muscle protein synthesis in old rats," Experimental Gerontology, vol. 39, no. 9, pp. 1315–1321, 2004.

9.   I. Rieu, H. Magne, I. Savary-Auzeloux et al., "Reduction of low grade inflammation restores blunting of postprandial muscle anabolism and limits sarcopenia in old rats," Journal of Physiology, vol. 587, no. 22, pp. 5483–5492, 2009.

10.   C. H. Lang and R. A. Frost, "Glucocorticoids and TNFα interact cooperatively to mediate sepsis-induced leucine resistance in skeletal muscle," Molecular Medicine, vol. 12, no. 11-12, pp. 291–299, 2006.

11.   B. Marzani, M. Balage, A. Vénien et al., "Antioxidant supplementation restores defective leucine stimulation of protein synthesis in skeletal muscle from old rats," Journal of Nutrition, vol. 138, no. 11, pp. 2205–2211, 2008.

12.   D. Dardevet, C. Sornet, G. Bayle, J. Prugnaud, C. Pouyet, and J. Grizard, "Postprandial stimulation of muscle protein synthesis in old rats can be restored by a leucine-supplemented meal," Journal of Nutrition, vol. 132, no. 1, pp. 95–100, 2002.

13.   C. S. Katsanos, H. Kobayashi, M. Sheffield-Moore, A. Aarsland, and R. R. Wolfe, "A high proportion of leucine is required for optimal stimulation of the rate of muscle protein synthesis by essential amino acids in the elderly," American Journal of Physiology, vol. 291, no. 2, pp. E381–E387, 2006.

14.   L. Combaret, D. Dardevet, I. Rieu et al., "A leucine-supplemented diet restores the defective postprandial inhibition of proteasome-dependent proteolysis in aged rat skeletal muscle," Journal of Physiology, vol. 569, no. 2, pp. 489–499, 2005.

15.   D. Dardevet, C. Sornet, M. Balage, and J. Grizard, "Stimulation of in vitro rat muscle protein synthesis by leucine decreases with age," Journal of Nutrition, vol. 130, no. 11, pp. 2630–2635, 2000.

16.   M. Balage and D. Dardevet, "Long-term effects of leucine supplementation on body composition," Current Opinion in Clinical Nutrition and Metabolic Care, vol. 13, no. 3, pp. 265–270, 2010.

17.   O. Pansarasa, V. Flati, G. Corsetti, L. Brocca, E. Pasini, and G. D'Antona, "Oral amino acid supplementation counteracts age-induced sarcopenia in elderly rats," American Journal of Cardiology, vol. 101, no. 11, pp. S35–S41, 2008.

18.   S. Verhoeven, K. Vanschoonbeek, L. B. Verdijk et al., "Long-term leucine supplementation does not increase muscle mass or strength in healthy elderly men," American Journal of Clinical Nutrition, vol. 89, no. 5, pp. 1468–1475, 2009.

19.   G. Zéanandin, M. Balage, S. M.. Schneider, J. Dupont, and I. Mothe-Satney, "Long-term leucine-enriched diet increases adipose tissue mass without affecting skeletal muscle mass and overall insulin sensitivity in old rats," Age, vol. 34, no. 2, pp. 371–387, 2012.

20.   M. Dangin, Y. Boirie, C. Guillet, and B. Beaufrère, "Influence of the protein digestion rate on protein turnover in young and elderly subjects," Journal of Nutrition, vol. 132, no. 10, pp. 3228S–3233S, 2002.

21.   I. Rieu, M. Balage, C. Sornet et al., "Increased availability of leucine with leucine-rich whey proteins improves postprandial muscle protein synthesis in aging rats," Nutrition, vol. 23, no. 4, pp. 323–331, 2007.

22.   H. Magne, I. Savary-Auzeloux, C. Migné et al., "Contrarily to whey and high protein diets, dietary free leucine supplementation cannot reverse the lack of recovery of muscle mass after prolonged immobilization during ageing," Journal of Physiology, vol. 590, pp. 2035–2049, 2012.

23. Y. Boirie, M. Dangin, P. Gachon, M. P. Vasson, J. L. Maubois, and B. Beaufrère, "Slow and fast dietary proteins differently modulate postprandial protein accretion," Proceedings of the National Academy of Sciences of the United States of America, vol. 94, no. 26, pp. 14930–14935, 1997.

24. M. A. Arnal, L. Mosoni, Y. Boirie et al., "Protein pulse feeding improves protein retention in elderly women," American Journal of Clinical Nutrition, vol. 69, no. 6, pp. 1202–1208, 1999.

25. M. A. Arnal, L. Mosoni, Y. Boirie et al., "Protein turnover modifications induced by the protein feeding pattern still persist after the end of the diets," American Journal of Physiology, vol. 278, no. 5, pp. E902–E909, 2000.

26. M. A. Arnal, L. Mosoni, D. Dardevet et al., "Pulse protein feeding pattern restores stimulation of muscle protein synthesis during the feeding period in old rats," Journal of Nutrition, vol. 132, no. 5, pp. 1002–1008, 2002.

27. R. Scognamiglio, A. Avogaro, C. Negut, R. Piccolotto, S. Vigili de Kreutzenberg, and A. Tiengo, "The effects of oral amino acid intake on ambulatory capacity in elderly subjects," Aging, vol. 16, no. 6, pp. 443–447, 2004.

28. R. Scognamiglio, R. Piccolotto, C. Negut, A. Tiengo, and A. Avogaro, "Oral amino acids in elderly subjects: effect on myocardial function and walking capacity," Gerontology, vol. 51, no. 5, pp. 302–308, 2005.

29. R. Scognamiglio, A. Testa, R. Aquilani, F. S. Dioguardi, and E. Pasini, "Impairment in walking capacity and myocardial function in the elderly: is there a role for non-pharmacologic therapy with nutritional amino acid supplements?" American Journal of Cardiology, vol. 101, no. 11, pp. S78–S81, 2008.

30. S. B. Solerte, C. Gazzaruso, R. Bonacasa et al., "Nutritional supplements with oral amino acid mixtures increases whole-body lean mass and insulin sensitivity in elderly subjects with sarcopenia," American Journal of Cardiology, vol. 101, no. 11, pp. S69–S77, 2008.

31. T. B. Harris, L. Ferrucci, R. P. Tracy et al., "Associations of elevated interleukin-6 and C-reactive protein levels with mortality in the elderly," American Journal of Medicine, vol. 106, no. 5, pp. 506–512, 1999.

32. I. Bautmans, R. Njemini, M. Lambert, C. Demanet, and T. Mets, "Circulating acute phase mediators and skeletal muscle performance in hospitalized geriatric patients," Journals of Gerontology Series A, vol. 60, no. 3, pp. 361–367, 2005.

33. L. A. Schaap, S. M. F. Pluijm, D. J. H. Deeg, and M. Visser, "Inflammatory markers and loss of muscle mass (Sarcopenia) and strength," American Journal of Medicine, vol. 119, no. 6, pp. 526.e9–526.e17, 2006.

34. M. Visser, M. Pahor, D. R. Taaffe et al., "Relationship of interleukin-6 and tumor necrosis factor-α with muscle mass and muscle strength in elderly men and women: the health ABC study," Journals of Gerontology Series A, vol. 57, no. 5, pp. M326–M332, 2002.

35. L. Mosoni, M. Balage, E. Vazeille et al., "Antioxidant supplementation had positive effects in old rat muscle, but through better oxidative status in other organs," Nutrition, vol. 26, no. 11-12, pp. 1157–1162, 2010.

36. G. I. Smith, P. Atherton, D. N. Reeds et al., "Dietary omega-3 fatty acid supplementation increases the rate of muscle protein synthesis in older adults: a randomized

controlled trial," American Journal of Clinical Nutrition, vol. 93, no. 2, pp. 402–412, 2011.

37. N. A. Burd, B. Y. Wall, and L. J. C. Van Loon, "The curious case of anabolic resistance: old wives' tales or new fables?" Journal of Applied Physiology, vol. 112, pp. 1233–1235, 2012.

CHAPTER 11

Rationale for Antioxidant Supplementation in Sarcopenia

FRANCESCO CERULLO, GIOVANNI GAMBASSI, AND MATTEO CESARI

11.1 INTRODUCTION

There is a common and diffuse false myth that aging is synonym of deterioration, pathology, and death. The increased life expectancy in developed and developing countries is parallel to the need to identify interventions able to preserve health and function even at older age, delaying the physical and cognitive declines. Aging is an extremely complex multifactorial process characterized by progressive physiological, genetic, and molecular changes, responsible for the increase risk of morbidity and death [1]. Several hypotheses [2, 3] have been proposed to explain this inborn process common to all living beings, but one of the most plausible and better-accepted currently is the so-called "free radical theory of aging."

Rationale for Antioxidant Supplementation in Sarcopenia. © Cerullo F, Gambassi G, and Cesari M. Journal of Aging Research *2012 (2012). http://dx.doi.org/10.1155/2012/316943. Licensed under a Creative Commons Attribution 3.0 Unported License, http://creativecommons.org/licenses/by/3.0/.*

The age-associated loss of skeletal muscle mass and strength (i.e., sarcopenia) seems an unavoidable part of the aging process. After about the age of 50 years, there is a progressive decrease of muscle mass at the rate of 1-2% per year. Similarly but with different decline rate and timing, muscle strength also decreases by about 3% yearly after 60 years of age [4]. Sarcopenia is a multidimensional phenomenon of aging (someone indicates it as a syndrome) and represents a powerful risk factor for the development of negative health-related events in the elderly. In fact, the relationships of sarcopenia with impaired physical performance, frailty, loss of functional independence, and increased risk of falls are all well established in the literature [5]. Moreover, decreased muscle strength is also highly predictive of incident disability and all-cause mortality in older persons [6].

Oxidative damage has been proposed as one of the major contributors of the skeletal muscle decline occurring with aging [7, 8]. The identification of free radicals as promoters of the aging process may imply that their inhibition might limit the detrimental modifications they exert on our organism (and, in particular, on skeletal muscle). In other words, if molecules with antioxidant capacities can counteract the oxidative damage, they may also play a key role in preventing the onset of age-related conditions, including the disabling process [9]. It will come to be true that oxidative damage is at the basis of the pathophysiological mechanisms responsible for sarcopenia (and other geriatric conditions), and interventions aimed at enhancing the endogenous antioxidant defenses (e.g., dietary antioxidant supplementation) may gain special interest.

The purpose of the present paper is to discuss current available evidence about the effects of antioxidant supplementation on sarcopenia. Special attention will be obviously given to studies focused on models of aging and involving older participants.

11.2 THE FREE RADICAL THEORY OF AGING

This theory was formulated for the first time by Harman in 1956 [10]. He proposed that aging and the associated degenerative diseases were consequences of free radical-induced damages to cells and the inability of counterbalancing these changes by endogenous antioxidant defenses. Harman initially explained the production of free radicals through reactions involving molecular oxygen catalyzed in cells by oxidative enzymes and subsequently postulated that genetic and environmental factors might modify this process. In 1972, he then revised his theory identifying the mitochondria as primarily responsible for the physiological process of aging [11]. Since oxidative damage is higher in cells and structures with higher consumption of oxygen, he suggested that mitochondria (consuming most of the intracellular oxygen) were particularly exposed to oxidative damage and potentially affected lifespan. Miquel and colleagues [12] subsequently confirmed such theory by recognizing mitochondria as major actors of cellular aging. More recently and consistently with these concepts, the free radical theory of aging has been switched into a "mitochondrial free radical theory of aging" [13].

Free radicals are a highly reactive chemical species with a single unpaired electron in its outer orbit seeking to pair with another free electron [14]. In particular, reactive oxygen species (ROS), deriving from oxidative metabolism, have higher reactivity than O_2. ROS are constantly generated in cells of aerobic organisms by the addition of a single electron to the oxygen molecule with subsequently damage of biological macromolecules (like lipids, proteins, and nucleic acids). The interaction of ROS with normal cellular structures leads to potentially nonreversible modifications, with consequent cellular loss of function and death [3, 15, 16].

There are numerous sites of oxidant generation [17]. Mitochondrial electron transport, peroxisomal fatty acid, cytochrome P-450, and phagocytic cells (the "respiratory burst") represent the most important ones. In particular, the main source of ROS (estimated at approximately 90% of the generated total) is located at the inner mitochondrial membrane where oxidative phosphorylation takes place [18]. Moreover, a variety of exogenous stimuli, such as exposure to infections [19], radiations [20], xenobiotics [21], environmental toxins [22], and ultraviolet light [23], may also increase the ROS production in vivo. Interestingly, mitochondria are both producers as well as targets of ROS.

In all organisms living in an aerobic environment, ROS play an important role in the maintenance of body homeostasis [24]. Recent studies [25, 26] have even shown that ROS may function as an additional class of cellular messengers, being involved as physiological regulators of intracellular signaling pathways (e.g., response to growth factor stimulation or generation of the inflammatory response to bacterial defense). Therefore, since free radicals are necessary for correct functioning of the human organism, efficient mechanisms of antioxidant defense had to be developed (especially in cells highly exposed to oxidation processes) with the aim of protecting cellular constituents. Endogenous antioxidants are present under diverse and numerous forms, like enzymes (e.g., superoxide dismutase, catalase, glutathione peroxidase, glutathione reductase), vitamins (e.g., vitamin C, vitamin E), and elements (e.g., selenium, zinc). All these substances are able to neutralize ROS and protect cells from free radical damage. Nevertheless, in a normal scenario, about 1% of ROS is still able to avoid the control of the antioxidant system, causing oxidation of surrounding tissues [27]. Moreover, oxidative damage is obviously enhanced when the ROS production increases and/or the antioxidant status decreases. According to the cellular constituents attacked by the oxidative stress, lipid peroxidation [28], protein oxidation [29], and DNA damage [30] will all promote abnormalities of the cell structures incompatible with proper cell function, leading to its death. Such cellular loss has been finally indicated as being responsible for the age-related degenerative diseases and conditions [31].

In summary, according to the free radical theory of aging, oxidative damage is due to a redox imbalance between ROS and the counteracting antioxidant forces generating a vicious cycle responsible for the progressive augmentation of damage [13]. The equilibrium between ROS production, antioxidant defenses, and the cellular structures determines whether a cell exposed to ROS increase will be destined to survival or death [32].

11.3 THE ANTIOXIDANT DEFENSE SYSTEM

Antioxidants are substances able to inhibit the rate of oxidation [33, 34]. Mainly, antioxidant enzymes (e.g., catalase, superoxide dismutase (SOD), glutathione peroxidase, glutathione reductase) work to maintain a state of balance preventing the transformation of ROS and to convert them into more stable molecules (like water and molecular oxygen). SOD exists in two forms: Cu/ZnSOD is present in the cytoplasm of the eukaryotic cells whilst MnSOD is located primarily in the mitochondria. Differently, endogenous nonenzymatic elements with antioxidant properties contribute to the maintenance of homeostasis by primarily acting as cofactors for the antioxidant enzymes. A major source of antioxidants is diet [35]. Among dietary antioxidants, the most important (and also largely available as supplements) are vitamin C, vitamin E, and carotenoids.

Vitamin C acts as an antioxidant by inhibiting oxidation. For example, ascorbic acid concentrations are inversely associated with isoprostanes, a marker of lipid peroxidation [36]. Vitamin C is involved in the regeneration of vitamin E in lipoproteins and membranes. In fact, vitamin C reduces the vitamin E radicals generated in cellular membranes and inhibits the propagation of α-tocopheroxyl radical. It is the most important hydrophilic antioxidant [27, 37, 38]. Conversely, vitamin E is the most important lipophilic antioxidant. It has been reported a significantly high correlation of α-tocopherol (the most diffuse form of circulating vitamin E) with physical performance and of γ-tocopherol with skeletal muscle strength [9]. Nunes and colleagues [39] demonstrated that a vitamin-E-deficient diet may cause an increased caspase-like activation (a proapoptotic stimulus) in animal model. Consistently, vitamin E is inversely correlated with lipid peroxidation and positively correlated with in cytochrome oxidase activity (causing a mitochondrial respiratory chain dysfunction) [40].

The supplementation of different antioxidants will provide different effects on oxidation according to the hydrophobicity of the administered molecule. In fact, fat-soluble antioxidants (e.g., vitamin E) are particularly effective at inhibiting lipoprotein peroxidation, whereas water-soluble antioxidants (e.g., vitamin C) more efficiently protect the aqueous phase. However, these antioxidants do not only act individually, but also cooperatively and sometimes even synergistically. Moreover, their radical-scavenging effects depend on various factors, the site of generation of the oxidant and the localization of the antioxidant [41, 42].

Another important source of antioxidants is assured by dietary carotenoids (including α-carotene, β-carotene, β-cryptoxanthin, lutein, zeaxanthin, and lycopene). Carotenoids are lipid soluble molecules that interact with the lipid membrane bilayer [43]. They scavenge free radicals, quench singlet oxygen, inhibit lipid peroxidation, and modulate redox-sensitive transcription factors (e.g., the nuclear factor kappa-light-chain-enhancer of activated B cells or NF-κB) that are involved in the upregulation of pro-inflammatory cytokines [44, 45]. Upritchard and colleagues [46] have reported that the supplementation of carotenoids may significantly improve antioxidant status and reduce concentrations of isoprostanes.

Numerous foods like fruits, nuts, vegetables, and spices are rich in antioxidants. Blueberries, cranberries, blackberries, plums, apples, cherries, and prunes are fruits particularly rich in antioxidants. Red and black beans, artichokes, and russet potatoes are the vegetables with highest content of antioxidants. Ground cloves, cinnamon, and oregano contain the greatest amount of antioxidants among spices [47].

Finally, it is worth to be mentioned that the antioxidant system function is highly influenced by age [32, 48]. In fact, with aging, there is a progressive decline in mitochondrial function and increase in oxidative damage. ROS overwhelms the endogenous antioxidant defense system during the aging process, causing harmful modifications of myofiber cellular proteins, lipids, and DNA. Moreover, the antioxidant dietary intake may easily decrease for multiple reasons (e.g., mobility disability leading to inadequate food supply; lack of social support; oral problems leading to repetitive and inadequate diets, etc.). Thus, it is not surprising that poor antioxidant status is commonly found at the foundations of several conditions of older age [49].

11.4 SARCOPENIA

In 1989, Rosenberg coined the term "sarcopenia" (from Greek: σαρξ or flesh and πενία or loss) to describe the progressive and involuntary loss of skeletal muscle mass and function occurring with aging [50]. Although it is difficult to provide an exact prevalence of sarcopenia due to the heterogeneity of older persons, the existing multiple operative definitions, and diverse (sometimes contradicting) assessment methodologies, common estimates indicate it as 5–13% in the 60–70-year-old age group and 11–50% in the 80-year and older age group [51]. Such figures are expected to dramatically rise in the next future due to the progressive aging of Western populations with a profound and negative impact on public health. In fact, sarcopenia (one of the most important geriatric syndromes) represents an important risk factor for functional impairment [52], physical disability [53, 54], falls [55], and frailty [56, 57].

Skeletal muscle is a tissue formed by multiple types of fibers. Briefly, type I fibers are slow contracting and use an oxidative metabolism. Differently, type II fibers are fast contracting and mainly glycolytic. Sarcopenic muscle mass reduction is primarily due to a loss of muscle fibers particularly characterized by a preferential atrophy of type II fibers. At the same time, a conversion of fast type II muscle fibers into slow type I fibers (with resulting loss in muscle power and decline in protein synthesis, especially of myosin heavy chains) has been described [58–60]. Overall, these changes lead to a smaller, slower contracting muscle with resulting reduced capacity to adequately perform activities of daily living. These anatomical modifications have been (at least partly) attributed to the age-related increase of oxidative damage. In fact, the skeletal muscle is the largest consumer of oxygen in the body with muscle fibers continuously generating ROS (especially during the contractile activity). Studies adopting muscle biopsies have confirmed that markers of oxidative damage are particularly and locally elevated in skeletal muscle of older adults [61–64], promoting the above-mentioned inadequacy of the antioxidant system in preventing damages [32, 65]. Some components of the enzymatic scavenger system, such as catalase, glutathione transferase, and superoxide dismutase, are also significantly depressed in elderly muscle. The consequent prooxidant status results in the alteration of mitochondrial DNA and abnormalities in the electron transport system, leading to reduced calcium uptake by the sarcoplasmic reticulum [66], irreversible damage of the cell, and its consequent death [67–69].

In the healthy muscle, proteins and amino acids are ideally balanced between synthesis and breakdown. In the elderly, this equilibrium is disrupted because of a lower synthesis and an increased breakdown rate of myofibrillar and mitochondrial proteins. Several endogenous and exogenous factors may affect the organism capacity to maintain the protein homeostasis. For example, it has been hypothesized that muscle decline might be caused by the direct detrimental effects of the chronic low-grade inflammatory status of advanced age. In this context, it cannot be ignored the close relationship between inflammation and oxidative damage. Interestingly, both inflammation (in particular through the TNF-α pathway) and oxidative damage are major regulator of cellular apoptosis and protein metabolism. At the same time, the age-related muscle protein loss may also macroscopically be caused by the frequent reduction of food intake in elders. Older persons are particularly exposed to the risk of (micro- and macronutrient) malnutrition that might also be related to endogenous (e.g., malabsorption, edentulism) or exogenous (e.g., lack of social support, disability) causes. Consequently, antioxidant supplementation represents a promising intervention potentially able to correct the inadequate diet of older persons and prevent the skeletal muscle decline by inhibiting the vicious cycle at the basis of protein catabolism.

Antioxidant supplementation may not replace the age-related decline of the more complex antioxidant enzymatic system. Nevertheless, in theory, by reinforcing the antioxidant nonenzymatic defences, supplementation might still be helpful in preventing the onset of age-related conditions (including sarcopenia) by acting on the same cause (i.e., oxidation). However, further studies are needed to better understand the relationship between nonenzymatic and enzymatic antioxidant defenses. In particular, it could be that the two components are not parallel and independent, but indeed synergistically constitute the antioxidant system. Interestingly, Selman and colleagues [70] recently demonstrated that vitamin C supplementation in mice is associated with a downregulation of antioxidant protection genes (including MnSOD).

11.5 ANTIOXIDANT SUPPLEMENTATION AND SARCOPENIA

Despite the clinical relevance of sarcopenia and the large interest for antioxidant supplementation (both from a research and commercial standpoint of view), evidence in this field is extremely limited and controversial. Most of the studies available in the literature are from epidemiological data with most even coming from cross-sectional observations.

The absence of a unique operative definition of sarcopenia represents a major issue limiting the conduction of research on the topic. In fact, most of the human studies analyze the relationship between dietary antioxidant supplementation and physical performance or muscle strength measures, without specifically focusing on the broader condition of sarcopenia. In fact, the quantitative (i.e., muscle mass) and qualitative (i.e., muscle strength, muscle function) declines follow separate and different trajectories with aging. The need of combining them into one single bidimensional definition of sarcopenia is critical from a theoretical and methodological perspective and has to be taken into account when specifically facing the topic sarcopenia. In other words, the separate evaluation of muscle mass or muscle strength may significantly affect the study of sarcopenia (limiting its exploration to only one of the two components) and likely result in biased findings. To our knowledge, there are currently no trials verifying the effects of antioxidant supplementation on sarcopenia (as identified by one of several the consensus definitions provided by international groups of experts). Interestingly, a recent statement from the Society on Sarcopenia, Cachexia, and Wasting Disease does not even mention antioxidant supplementation as a possible tool to manage sarcopenia in older persons [71].

Several studies have been conducted to evaluate the effects of antioxidant supplementation on antioxidant status [72, 73]. Overall, results consistently report significant improvements of antioxidant biomarkers after a period of specific supplementation.

Differently, the effect of antioxidant supplementation on muscle performance is still and largely controversial. Here are just a few examples of positive studies (from both animal and human models) among the large body of literature suggesting a beneficial effect of antioxidant supplementation and sarcopenia. Jakeman and Maxwell [74] showed a protective effect of vitamin C supplementation against exercise-induced muscle damage. Similarly, Shafat and colleagues [75] reported a reduction of muscle damage by adopting a counteracting vitamin E and C supplementation. An antioxidant mixture containing vitamin E, vitamin A, rutin, zinc, selenium has shown to increase the anabolic response of old muscle to leucine and the leucine-induced inhibition of protein degradation in rats [76]. Resveratrol, a natural polyphenol found in grapes, peanuts, and berries [77], has shown a protective effect against oxidative stress in skeletal muscle through the expression of antioxidant enzymes [78]. At the same time, a similar (possibly larger) number of studies have reported negative results on the topic. For example, some trials have reported that antioxidant supplementation may even be unfavorable on physical performance and underlying biological mechanisms [79, 80]. A decrease of baseline levels of antioxidant mitochondrial enzymes was reported by Strobel and colleagues [81] after 14 weeks of vitamin E and α-lipoic acid administration in rats. Data by Ristow and colleagues [82] suggested that oral administration of ascorbic acid and α-tocopherol prevents exercise training-induced increases in insulin sensitivity and ROS defense capacity. Consistently, the work by Higashida and colleagues [83] showed no inhibitory effect on the adaptive responses of muscle to exercise by antioxidant vitamin supplementation. Several studies [84–86] showed no effects of antioxidants supplementation on muscle function after exercise-induced muscle damage. In a recent work, Theodorou and colleagues [87] found that a supplementation with ascorbic acid and tocopherol does not affect muscle performance. Kondratov and colleagues [88] investigated possible improvements of signs of premature aging (e.g., cataracts, cornea inflammation, joint ossification, and muscle waste) in BMAL1 knockout mice by supplementation of N-acetyl-L-cysteine (NAC), a low-molecular-weight antioxidant. Result suggested an extended lifespan, but excluded significant effects on sarcopenia. Sacheck et al. [89] showed α-tocopherol supplementation in older fit men did not suppress postexercise elevations in biomarkers of muscle damage and lipid peroxidation such as in younger men. More recently, vitamin E and C supplementation have shown to reduce muscular levels of oxidative stress in repetitively loaded muscles of old rats, but no increase in muscle mass and maximal force production (after more than 4 weeks of training) was found [90]. Consistent results were obtained by Barker and colleagues [91] after administration of vitamin E and C in men undergoing anterior cruciate ligament surgical repair. It might seem that the biological effects of antioxidant supplementations are easily captured as increased levels of antioxidant biomarkers. Differently, the clinically relevant and beneficial effects of antioxidant supplementation are more difficult to be obtained (always if obtainable!).

Possible reasons for these controversial results are various. First of all, it is possible that sarcopenia is not related to oxidative damage, so that the obtained negative results are indeed "true negative" findings. Second, it is not automatic that the modification of a biomarker concentrations is able to parallely change clinical parameters. It is more likely that subclinical effects are more sensible to changes than clinically evident manifestations. And this is particularly true when testing interventions in an extremely complex field as geriatric syndromes, in which one sign/symptom is not necessarily indicative of a well-defined condition. Different types and doses of antioxidants administered, timing of supplementation, the adopted animal species, and the measured biomarkers may represent other important causes of the negative findings. As mentioned, the heterogeneity and methodological limitations affecting the study of sarcopenia may further explain the controversial findings.

Finally, it is noteworthy the extreme scarcity of available clinical trials in humans on this topic. In fact, most of the positive results are obtained in animal models and still wait to be confirmed in humans. Furthermore, recent studies [92] reporting possible negative effects (e.g., cardiovascular and all-cause mortality) of long-term, high-dosage antioxidant supplementation (in particular, for vitamin E) cannot be ignored.

In the present paper, we chose not to go into many details about posologies of antioxidant supplementations. Current literature on the topic is extremely heterogeneous for methodological approaches, study designs (i.e., in vivo, in vitro, epidemiological evaluations, clinical trials, animal or human models), interventions (single antioxidant molecules or in combination, timing of administration, posologies), and outcome measures (i.e., biomarkers of muscle decline). Such heterogeneity is not surprising considering the relative novelty of the topic. After all, sarcopenia is a condition theoretically defined only about 20 years ago, but its operative definition is still debated.

11.6 CONCLUSION

In summary, there is some evidence that oral antioxidant supplementation may reduce muscle damage, but experimental results are largely preliminary and far to be clinically relevant, at least, as suggestive of positive benefits. In fact, a large body of evidence may indicate extreme cautiousness in taking antioxidant supplementation as preventive measures against aging process and age-related conditions. Further studies are needed to support the widespread practice of oral antioxidant supplementation and to determine appropriate recommendations in elderly.

Although antioxidant supplementation through diet is receiving growing attention, supporting evidence is still scarce and equivocal. Antioxidant supplementation could benefit muscle protein metabolism during aging, but further trials in humans and with adequate sample sizes are required to clearly establish the hypothesized relationship between antioxidants and sarcopenia. In this context, a better understanding of oxidation mechanisms, timing and doses of antioxidant supplementation, and appropriate methodological approaches to study this theme is needed to provide convincing evidence and justify the current widespread use of antioxidants supplementation.

REFERENCES

1. D. Harman, "The free radical theory of aging," Antioxid Redox Signal, vol. 5, no. 5, pp. 557–561, 2003.
2. Z. A. Medvedev, "An attempt at a rational classification of theories of ageing," Biological Reviews of The Cambridge Philosophical Society, vol. 65, no. 3, pp. 375–398, 1990.
3. B. T. Weinert and P. S. Timiras, "Theories of aging," Journal of Applied Physiology, vol. 95, no. 4, pp. 1706–1716, 2003.
4. S. von Haehling, J. E. Morley, and S. D. Anker, "An overview of sarcopenia: facts and numbers on prevalence and clinical impact," The Journal of Cachexia, Sarcopenia and Muscle, vol. 1, no. 2, pp. 129–133, 2010.
5. R. Roubenoff, "Sarcopenia and its implications for the elderly," European Journal of Clinical Nutrition, vol. 54, no. 3, pp. S40–S47, 2000.
6. E. J. Metter, L. A. Talbot, M. Schrager, and R. Conwit, "Skeletal muscle strength as a predictor of all-cause mortality in healthy men," The Journals of Gerontology Series A, vol. 57, no. 10, pp. B359–B365, 2002.
7. O. Pansarasa, L. Castagna, B. Colombi, J. Vecchiet, G. Felzani, and F. Marzatico, "Age and sex differences in human skeletal muscle: role of reactive oxygen species," Free Radical Research, vol. 33, no. 3, pp. 287–293, 2000.
8. G. Fanò, P. Mecocci, J. Vecchiet, et al., "Age and sex influence on oxidative damage and functional status in human skeletal muscle," Journal of Muscle Research and Cell Motility, vol. 22, no. 4, pp. 345–351, 2001.
9. M. Cesari, M. Pahor, B. Bartali, et al., "Antioxidants and physical performance in elderly persons: the Invecchiare in Chianti (InCHIANTI) study," The American Journal of Clinical Nutrition, vol. 79, no. 2, pp. 289–294, 2004.
10. D. Harman, "Aging: a theory based on free radical and radiation chemistry," The Journals of Gerontology, vol. 11, no. 3, pp. 298–300, 1956.

11. D. Harman, "The biologic clock: the mitochondria?" The Journal of The American Geriatrics Society, vol. 20, no. 4, pp. 145–147, 1972.

12. J. Miquel, A. C. Economos, J. Fleming, and J. E. Johnson Jr., "Mitochondrial role in cell aging," Experimental Gerontology, vol. 15, no. 6, pp. 575–591, 1980.

13. J. Sastre, F. V. Pallardó, J. García de la Asunción, and J. Viña, "Mitochondria, oxidative stress and aging," Free Radical Research, vol. 32, no. 3, pp. 189–198, 2000.

14. B. H. Halliwell and J. M. C. Gutteridge, Free Radicals in Biology and Medicine, Oxford University Press, Oxford, UK, 1989.

15. K. B. Beckman and B. N. Ames, "The free radical theory of aging matures," Physiological Reviews, vol. 78, no. 2, pp. 547–581, 1998.

16. P. Rossi, B. Marzani, S. Giardina, M. Negro, and F. Marzatico, "Human skeletal muscle aging and the oxidative system: cellular events," Current Aging Science, vol. 1, no. 3, pp. 182–191, 2008.

17. L. Gil Del Valle, "Oxidative stress in aging: theoretical outcomes and clinical evidences in humans," Biomedicine & Pharmacotherapy. In press.

18. R. S. Balaban, S. Nemoto, and T. Finkel, "Mitochondria, oxidants, and aging," Cell, vol. 120, no. 4, pp. 483–495, 2005.

19. K. B. Schwarz, "Oxidative stress during viral infection: a review," Free Radical Biology & Medicine, vol. 21, no. 5, pp. 641–649, 1996.

20. P. A. Riley, "Free radicals in biology: oxidative stress and the effects of ionizing radiation," International Journal of Radiation Biology, vol. 65, no. 1, pp. 27–33, 1994.

21. G. Pagano, "Redox-modulated xenobiotic action and ROS formation: a mirror or a window?" Human & Experimental Toxicology, vol. 21, no. 2, pp. 77–81, 2002.

22. H. Shi, X. Shi, and K. Liu, "Oxidative mechanism of arsenic toxicity and carcinogenesis," Molecular and Cellular Biochemistry, vol. 255, no. 1-2, pp. 67–78, 2004.

23. K. Scharffetter-Kochanek, M. Wlaschek, P. Brenneisen, M. Schauen, R. Blaudschun, and J. Wenk, "UV-induced reactive oxygen species in photocarcinogenesis and photoaging," The Journal of Biological Chemistry, vol. 378, no. 11, pp. 1247–1257, 1997.

24. C. Gemma, J. Vila, A. Bachstetter, and P. C. Bickford, "Oxidative stress and the aging brain: from theory to prevention," in Brain Aging: Models, Methods, and Mechanisms, D. R. Riddle, Ed., chapter 15, CRC Press, Boca Raton, Fla, USA, 2007.

25. L. T. Knapp and E. Klann, "Role of reactive oxygen species in hippocampal long-term potentiation: contributory or inhibitory?" Journal of Neuroscience Research, vol. 70, no. 1, pp. 1–7, 2002.

26. T. Finkel, "Signal transduction by reactive oxygen species," Journal of Cell Biology, vol. 194, no. 1, pp. 7–15, 2011.

27. D. Fusco, G. Colloca, M. R. Lo Monaco, and M. Cesari, "Effects of antioxidant supplementation on the aging process," Journal of Clinical Interventions in Aging, vol. 2, no. 3, pp. 377–387, 2007.

28. F. Q. Schafer and G. R. Buettner, "Acidic pH amplifies iron-mediated lipid peroxidation in cells," Free Radical Biology & Medicine, vol. 28, no. 8, pp. 1175–1181, 2000.

29. E. R. Stadtman, "Role of oxidant species in aging," Current Medicinal Chemistry, vol. 11, no. 9, pp. 1105–1112, 2004.

30. I. S. Kil, T. L. Huh, Y. S. Lee, Y. M. Lee, and J. W. Park, "Regulation of replicative senescence by NADP+-dependent isocitrate dehydrogenase," Free Radical Biology & Medicine, vol. 40, no. 1, pp. 110–119, 2006.

31. A. Goswami, P. Dikshit, A. Mishra, S. Mulherkar, N. Nukina, and N. R. Jana, "Oxidative stress promotes mutant huntingtin aggregation and mutant huntingtin-dependent cell death by mimicking proteasomal malfunction," Biochemical and Biophysical Research Communications, vol. 342, no. 1, pp. 184–190, 2006.

32. K. C. Kregel and H. J. Zhang, "An integrated view of oxidative stress in aging: basic mechanisms, functional effects, and pathological considerations," American Journal of Physiology—Regulatory, Integrative and Comparative Physiology, vol. 292, no. 1, pp. R18–R36, 2007.

33. D. Trachootham, W. Lu, MA. Ogasawara, et al., "Redox regulation of cell survival," Antioxidants & Redox Signaling, vol. 10, no. 8, pp. 1343–1374, 2008.

34. P. M. Clarkson and S. Thompson, "Antioxidants: what role do they play in physical activity and health?" American Journal of Clinical Nutrition, vol. 72, supplement 2, pp. 637S–646S, 2000.

35. M. M. Berger, "Can oxidative damage be treated nutritionally?" Clinical Nutrition, vol. 24, no. 2, pp. 172–183, 2005.

36. G. Block, M. Dietrich, E. P. Norkus, et al., "Factors associated with oxidative stress in human populations," American Journal of Epidemiology, vol. 156, no. 3, pp. 274–285, 2002.

37. E. Niki, Y. Yamamoto, M. Takahashi, et al., "Free radical-mediated damage of blood and its inhibition by antioxidants," Journal of Nutritional Science and Vitaminology, vol. 34, no. 5, pp. 507–512, 1988.

38. B. Frei, L. England, and B. N. Ames, "Ascorbate is an outstanding antioxidant in human blood plasma," Proceedings of the National Academy of Sciences of the United States of America, vol. 86, no. 16, pp. 6377–6381, 1989.

39. V. A. Nunes, A. J. Gozzo, M. A. Juliano, et al., "Antioxidant dietary deficiency induces caspase activation in chick skeletal muscle cells," Brazilian Journal of Medical and Biological Research, vol. 36, no. 8, pp. 1047–1053, 2003.

40. R. Rafique, A. H. Schapira, and J. M. Coper, "Mitochondrial respiratory chain dysfunction in ageing; influence of vitamin E deficiency," Free Radical Research, vol. 38, no. 2, pp. 157–165, 2004.

41. E. Niki, N. Noguchi, H. Tsuchihashi, and N. Gotoh, "Interaction among vitamin C, vitamin E, and beta-carotene," American Journal of Clinical Nutrition, vol. 62, no. 6, pp. 1322S–1326S, 1995.

42. T. Rinne, E. Mutschler, G. Wimmer-Greinecker, A. Moritz, and H. G. Olbrich, "Vitamins C and E protect isolated cardiomyocytes against oxidative damage," International Journal of Cardiology, vol. 75, no. 2-3, pp. 275–281, 2000.

43. D. Semba, F. Lauretani, and L. Ferrucci, "Carotenoids as protection against sarcopenia in older adults," Archives of Biochemistry and Biophysics, vol. 458, no. 2, pp. 141–145, 2007.

44. P. Hu, D. B. Reuben, E. M. Crimmins, T. B. Harris, M. H. Huang, and T. E. Seeman, "The effects of serum beta-carotene concentration and burden of inflammation on all-cause mortality risk in high-functioning older persons: MacArthur studies of suc-

cessful aging," The Journals of Gerontology Series A, vol. 59, no. 8, pp. 849–854, 2004.

45. J. Walston, Q. Xue, R. D. Semba, et al., "Serum antioxidants, inflammation, and total mortality in older women," American Journal of Epidemiology, vol. 163, no. 1, pp. 18–26, 2006.

46. J. E. Upritchard, C. R. Schuurman, A. Wiersma, et al., "Spread supplemented with moderate doses of vitamin E and carotenoids reduces lipid peroxidation in healthy, nonsmoking adults," American Journal of Clinical Nutrition, vol. 78, no. 5, pp. 985–992, 2003.

47. X. Wu, G. R. Beecher, J. M. Holden, D. B. Haytowitz, S. E. Gebhardt, and R. L. Prior, "Lipophilic and hydrophilic antioxidant capacities of common foods in the United States," Journal of Agricultural and Food Chemistry, vol. 52, no. 12, pp. 4026–4037, 2004.

48. K. F. Petersen, D. Befroy, S. Dufour, et al., "Mitochondrial dysfunction in the elderly: possible role in insulin resistance," Science, vol. 300, no. 5622, pp. 1140–1142, 2003.

49. S. R. J. Maxwell, "Prospects for the use of antioxidant therapies," Drugs, vol. 49, no. 3, pp. 345–361, 1995.

50. I. H. Rosenberg, "Sarcopenia: origins and clinical relevance," Journal of Nutrition, vol. 127, no. 5, supplement, pp. S990–S991, 1997.

51. D. L. Waters, R. N. Baumgartner, P. J. Garry, and B. Vellas, "Advantages of dietary, exercise-related, and therapeutic interventions to prevent and treat sarcopenia in adult patients: an update," Journal of Clinical Interventions in Aging, vol. 5, pp. 259–270, 2010.

52. I. Janssen, S. B. Heymsfield, and R. Ross, "Low relative skeletal muscle mass (sarcopenia) in older persons is associated with functional impairment and physical disability," Journal of the American Geriatrics Society, vol. 50, no. 5, pp. 889–896, 2002.

53. R. N. Baumgartner, K. M. Koehler, D. Gallagher, et al., "Epidemiology of sarcopenia among the elderly in New Mexico," American Journal of Epidemiology, vol. 147, no. 8, pp. 755–763, 1998.

54. I. Janssen, D. S. Shepard, P. T. Katzmarzyk, and R. Roubenoff, "The healthcare costs of sarcopenia in the United States," Journal of the American Geriatrics Society, vol. 52, no. 1, pp. 80–85, 2004.

55. L. Wolfson, J. Judge, R. Whipple, and M. King, "Strength is a major factor in balance, gait, and the occurrence of falls," The Journals of Gerontology Series A, vol. 50, pp. 64–67, 1995.

56. T. B. Vanitallie, "Frailty in the elderly: contributions of sarcopenia and visceral protein depletion," Metabolism, vol. 52, no. 10, supplement 2, pp. 22–26, 2003.

57. J. M. Bauer and C. C. Sieber, "Sarcopenia and frailty: a clinician's controversial point of view," Experimental Gerontology, vol. 43, no. 7, pp. 674–678, 2008.

58. H. Akima, Y. Kano, Y. Enomoto, et al., "Muscle function in 164 men and women aged 20–84 yr," Medicine & Science in Sports & Exercise, vol. 33, no. 2, pp. 220–226, 2001.

59. L. A. Burton and D. Sumukadas, "Optimal management of sarcopenia," Journal of Clinical Interventions in Aging, vol. 5, pp. 217–228, 2010.

60. J. E. Morley, R. N. Baumgartner, R. Roubenoff, J. Mayer, and K. S. Nair, "Sarco-penia," Journal of Laboratory and Clinical Medicine, vol. 137, no. 4, pp. 231–243, 2001.

61. O. Pansarasa, L. Bertorelli, J. Vecchiet, G. Felzani, and F. Marzatico, "Age-dependent changes of antioxidant activities and markers of free radical damage in human skeletal muscle," Free Radical Biology and Medicine, vol. 27, no. 5-6, pp. 617–622, 1999.

62. P. Mecocci, G. Fanó, S. Fulle, et al., "Age-dependent increases in oxidative damage to DNA, lipids, and proteins in human skeletal muscle," Free Radical Biology & Medicine, vol. 26, no. 3-4, pp. 303–308, 1999.

63. P. S. Lim, Y. M. Cheng, and Y. H. Wei, "Increase in oxidative damage to lipids and proteins in skeletal muscle of uremic patients," Free Radical Research, vol. 36, no. 3, pp. 295–301, 2002.

64. P. Gianni, K. J. Jan, M. J. Douglas, P. M. Stuart, and M. A. Tarnopolsky, "Oxidative stress and the mitochondrial theory of aging in human skeletal muscle," Experimental Gerontology, vol. 39, no. 9, pp. 1391–1400, 2004.

65. J. Palomero and M. J. Jackson, "Redox regulation in skeletal muscle during contractile activity and aging," Journal of Animal Science, vol. 88, no. 4, pp. 1307–1313, 2010.

66. S. Fulle, F. Protasi, G. Di Tano, et al., "The contribution of reactive oxygen species to sarcopenia and muscle ageing," Experimental Gerontology, vol. 39, no. 1, pp. 17–24, 2004.

67. M. B. Reid and Y. P. Li, "Cytokines and oxidative signalling in skeletal muscle," Acta Physiologica Scandinavica, vol. 171, no. 3, pp. 225–232, 2001.

68. A. Dirks and C. Leeuwenburgh, "Apoptosis in skeletal muscle with aging," American Journal of Physiology—Regulatory, Integrative and Comparative Physiology, vol. 282, no. 2, pp. R519–R527, 2002.

69. J. S. Kim, J. M. Wilson, and S. R. Lee, "Dietary implications on mechanisms of sarcopenia: roles of protein, amino acids and antioxidants," The Journal of Nutritional Biochemistry, vol. 21, no. 1, pp. 1–13, 2010.

70. C. Selman, J. S. McLaren, C. Meyer, et al., "Life-long vitamin C supplementation in combination with cold exposure does not affect oxidative damage or lifespan in mice, but decreases expression of antioxidant protection genes," Mechanisms of Ageing and Development, vol. 127, no. 12, pp. 897–904, 2006.

71. J. E. Morley, A. M. Abbatecola, J. M. Argiles, et al., "Sarcopenia with limited mobility: an international consensus," Journal of the American Medical Directors Association, vol. 12, no. 6, pp. 403–409, 2011.

72. A. D. Gupta, S. A. Dhundasi, J. G. Ambekar, and K. K. Das, "Effect of l-ascorbic acid on antioxidant defense system in testes of albino rats exposed to nickel sulfate," Journal of Basic and Clinical Physiology and Pharmacology, vol. 18, no. 4, pp. 255–266, 2007.

73. R. Rodrigo, H. Prat, W. Passalacqua, J. Araya, and J. P. Bächler, "Decrease in oxidative stress through supplementation of vitamins C and E is associated with a reduction in blood pressure in patients with essential hypertension," Clinical Science, vol. 114, no. 10, pp. 625–634, 2008.

74. P. Jakeman and S. Maxwell, "Effect of antioxidant vitamin supplementation on muscle function after eccentric exercise," European Journal of Applied Physiology and Occupational Physiology, vol. 67, no. 5, pp. 426–430, 1993.

75. A. Shafat, P. Butler, R. L. Jensen, and A. E. Donnelly, "Effects of dietary supplementation with vitamins C and E on muscle function during and after eccentric contractions in humans," European Journal of Applied Physiology, vol. 93, no. 1-2, pp. 196–202, 2004.

76. B. Marzani, M. Balage, A. Vénien, et al., "Antioxidant supplementation restores defective leucine stimulation of protein synthesis in skeletal muscle from old rats," Journal of Nutrition, vol. 138, no. 11, pp. 2205–2211, 2008.

77. J. A. Baur and D. A. Sinclair, "Therapeutic potential of resveratrol: the in vivo evidence," Nature Reviews Drug Discovery, vol. 5, no. 6, pp. 493–506, 2006.

78. J. R. Jackson, M. J. Ryan, Y. Hao, and S. E. Alway, "Mediation of endogenous antioxidant enzymes and apoptotic signaling by resveratrol following muscle disuse in the gastrocnemius muscles of young and old rats," American Journal of Physiology—Regulatory, Integrative and Comparative Physiology, vol. 299, no. 6, pp. R1572–R1581, 2010.

79. R. J. Marshall, K. C. Scott, R. C. Hill, et al., "Supplemental vitamin C appears to slow racing greyhounds," Journal of Nutrition, vol. 132, no. 6, pp. 1616S–1621S, 2002.

80. M. C. Gomez-Cabrera, E. Domenech, M. Romagnoli, et al., "Oral administration of vitamin C decreases muscle mitochondrial biogenesis and hampers training-induced adaptations in endurance performance," American Journal of Clinical Nutrition, vol. 87, no. 1, pp. 142–149, 2008.

81. N. A. Strobel, J. M. Peake, A. Matsumoto, S. A. Marsh, J. S. Coombes, and G. D. Wadley, "Antioxidant supplementation reduces skeletal muscle mitochondrial biogenesis," Medicine & Science in Sports & Exercise, vol. 43, no. 6, pp. 1017–1024, 2011.

82. M. Ristow, K. Zarse, A. Oberbach, et al., "Antioxidants prevent health-promoting effects of physical exercise in humans," Proceedings of the National Academy of Sciences of the United States of America, vol. 106, no. 21, pp. 8665–8670, 2009.

83. K. Higashida, S. H. Kim, M. Higuchi, J. O. Holloszy, and D. H. Han, "Normal adaptations to exercise despite protection against oxidative stress," American Journal of Physiology—Endocrinology and Metabolism, vol. 301, no. 5, pp. E779–E784, 2011.

84. D. M. Bailey, C. Williams, J. A. Betts, D. Thompson, and T. L. Hurst, "Oxidative stress, inflammation and recovery of muscle function after damaging exercise: effect of 6-week mixed antioxidant supplementation," European Journal of Applied Physiology, vol. 111, no. 6, pp. 925–936, 2011.

85. G. L. Close, T. Ashton, T. Cable, et al., "Ascorbic acid supplementation does not attenuate post-exercise muscle soreness following muscle-damaging exercise but may delay the recovery process," British Journal of Nutrition, vol. 95, no. 5, pp. 976–981, 2006.

86. E. W. Petersen, K. Ostrowski, T. Ibfelt, et al., "Effect of vitamin supplementation on cytokine response and on muscle damage after strenuous exercise," American Journal of Physiology—Cell Physiology, vol. 280, no. 6, pp. C1570–C1575, 2001.

87. A. A. Theodorou, M. G. Nikolaidis, V. Paschalis, et al., "No effect of antioxidant supplementation on muscle performance and blood redox status adaptations to eccentric training," The American Journal of Clinical Nutrition, vol. 93, no. 6, pp. 1373–1383, 2011.

88. R. V. Kondratov, O. Vykhovanets, A. A. Kondratova, and M. P. Antoch, "Antioxidant N-acetyl-L-cysteine ameliorates symptoms of premature aging associated with the deficiency of the circadian protein BMAL1," Aging, vol. 1, no. 12, pp. 979–987, 2009.

89. J. M. Sacheck, P. E. Milbury, J. G. Cannon, R. Roubenoff, and J. B. Blumberg, "Effect of vitamin E and eccentric exercise on selected biomarkers of oxidative stress in young and elderly men," Free Radical Biology and Medicine, vol. 34, no. 12, pp. 1575–1588, 2003.

90. M. J. Ryan, H. J. Dudash, M. Docherty, et al., "Vitamin E and C supplementation reduces oxidative stress, improves antioxidant enzymes and positive muscle work in chronically loaded muscles of aged rats," Experimental Gerontology, vol. 45, no. 11, pp. 882–895, 2010.

91. T. Barker, S. W. Leonard, J. Hansen, et al., "Vitamin E and C supplementation does not ameliorate muscle dysfunction after anterior cruciate ligament surgery," Free Radical Biology and Medicine, vol. 47, no. 11, pp. 1611–1618, 2009.

92. P. T. Gee, "Unleashing the untold and misunderstood observations on vitamin E," Genes & Nutrition, vol. 6, no. 1, pp. 5–16, 2011.

CHAPTER 12

Sarcopenia and Androgens: A Link Between Pathology and Treatment

CARLA BASUALTO-ALARCÓN, DIEGO VARELA, JAVIER DURAN, RODRIGO MAASS, AND MANUEL ESTRADA

12.1 INTRODUCTION

In healthy young adults, the skeletal muscle mass comprises approximately 60% of total body mass. From age 40, the percentage begins to decline, reaching 40% at 70 years-old (1). This age-related decline in skeletal muscle mass and strength generation, the primary function of skeletal muscle mass, is known as sarcopenia (2, 3). This newly identified syndrome impacts both quality and quantity of life for both men and women, often leading to physical disabilities, gait abnormalities, and falls that cause loss of functional independence (4). Although sarcopenia is affecting more and more population worldwide, its pathophysiology remains unclear. Such lack of clarity can be attributed to difficulty in isolating the individual events responsible for alterations in skeletal muscle, most of which occur simultaneously, among the multiple age-associated changes and co-mor-

Sarcopenia and Androgens: A Link Between Pathology and Rreatment. © *Basualto-Alarcón C, Varela D, Duran J, Maass R, and Estrada M.* Frontiers in Endocrinology *5,217 (2014), doi: 10.3389/ fendo.2014.00217. Licensed under a Creative Commons Attribution 4.0 International License, http:// creativecommons.org/licenses/by/4.0/.*

bidities associated with advanced age. Indeed, most of the intrinsic as well as extrinsic (systemic) muscle changes that occur with age are believed to be involved in the development of sarcopenia (5, 6).

The difficulty in defining sarcopenia has created challenges in determining the best treatment for patients with this disease. Among the many different therapeutic approaches, including exercise, hormone, nutritional therapy, and/or a combination thereof, no one approach has emerged as superior in animal or human studies, and the great variability in the clinical outcomes of patients treated with these approaches has prevented identification of the most effective therapy (7, 8). One approach that has drawn recent attention is supplementation with androgens, hormones with anabolic properties whose levels naturally decline with age (9–12). Clinical studies of androgen supplementation in age-related diseases and muscle wasting are a focus of emerging interest (11). Skeletal muscle is a target tissue for anabolic steroids. Testosterone elicits significant muscular effects and abnormalities of plasma concentrations can cause muscle disease (13). High levels are associated with muscle hypertrophy, whereas low levels are epidemiologically associated with metabolic syndrome and diabetes, which negatively impact muscle functions. However, most evidence linked testosterone and skeletal muscle effects are observational. Studies targeted at establishing such effects at cellular level and their correlations with in vivo models, will broaden our understanding of the roles played by androgens on skeletal muscle function in elderly.

12.2 PREVALENCE OF SARCOPENIA

The term sarcopenia was first proposed in 1989 by Irwin Rosenberg to describe a multifactorial syndrome that occurs with age and results in loss of skeletal muscle mass and function (3, 4). Prior to publication of the Report of the *European Working Group on Sarcopenia in Older People* (EWG-SOP) in 2010, a variety of definitions of sarcopenia and inclusion criteria for patients with sarcopenia were used in epidemiological studies. In this report, the EWGSOP defined sarcopenia as a syndrome characterized by progressive and generalized loss of skeletal muscle mass and strength associated with physical disability, poor quality of life, and risk of death, and they also recommended criteria for diagnosis (14). Importantly, the recent development of these clear definitions and guidelines to diagnose and treat sarcopenia has significantly improved understanding of this relatively newly identified condition.

Although epidemiological research into sarcopenia has identified factors potentially associated with its prevalence, such as testosterone levels (15), most research has been cross-sectional rather than longitudinal, the approach that is necessary to increase our understanding. The prevalence of sarcopenia is partially dependent on the population studied, the measurement technique employed, and the operational definition used. Some authors report that sarcopenia affects both Hispanic and non-Hispanic white men and women, with prevalence ranging from 13.5 to 24% in individuals under 70 years, reaching 60% in individuals above 80 years. Also sarcopenia was significantly associated with several co-morbidities and lifestyle factors, including obesity, low income, current smoking, and lung disease (4). Regarding gender differences in muscle distribution, it has been observed that woman had up to 40% less muscle in the upper body but no more than 33% less mass in the lower body compared to man (16). At a transcriptional level, a marked sexual dimorphism was observed on the biceps brachii muscle. According to the authors, it appears that mainly age-induced changes at the transcriptional level in women, were responsible for increased expression levels of signaling pathways dealing with muscle dysfunction, inflammation, and mitochondrial dysfunction, thus leading to an alteration in "muscle quality" (17). On the contrary, regarding total muscle mass, age-related loss in total body muscle mass was greater in the male than female subjects, independent of change in stature, and can be clearly observed after approximately age 45. Also these results indicate that age-related loss of skeletal muscle mass is greater in the lower than upper body in both men and women (16). Clinical studies reveal that sarcopenia is a main cause of higher frailty, impairment, and loss of independence in elderly women than men (18). In addition, other cause contributing to sarcopenia is related to physical activity, which declines with age. Other indirect factors such as inflammation and oxidative stress also have been suggested to contribute to sarcopenia development. One of these factors that represent differences in men and women is the etiology of adipose tissue, which is a potent source of pro-inflammatory cytokines, suggesting that sex differences in total and regional adiposity can impact sarcopenia development.

Observation of a relationship between testosterone level and skeletal muscle anabolism has led to research evaluating age-related declines in testosterone levels. In a cross-sectional study, Feldman et al. (19) determined that total testosterone level declined approximately 0.8% per year and that both free and albumin-bound testosterone level declined approximately 2% per year after age 40, whereas sex hormone-binding globulin (SHBG) increased by 1.6% per year. In a subsequent longitudinal analysis, they determined that total testosterone declined by 1.6% per year and bioavailable testosterone declined by 2–3% per year. In a third study, they found that both testosterone level and the free testosterone index decreased progressively from the third to the ninth decades of life, specifically that total testosterone level decreased an average 0.110 nmol/L per year and the free testosterone index an average of 0.0049 nmol testosterone per nanomole SHBG per year, regardless of whether these variables were examined cross-sectional or longitudinally (19).

12.3 CELLULAR AND MOLECULAR MECHANISMS UNDERLYING SARCOPENIA

Although the pathophysiology of sarcopenia remains under investigation, two main groups of factors have been identified as responsible for its development: (1) intrinsic factors at the cellular level, such as mitochondrial malfunction and oxidative stress; (2) systemic factors, such as production of inflammatory cytokines, changes in hormone levels, and overall decrease in physical activity level. Loss of muscle mass and functionality related to sarcopenia cannot be completely explained by atrophy of muscle fibers. As such, the involvement of several atrophy-independent mechanisms in these conditions has been postulated, including decrease in myosin force and/or actin–myosin cross-bridge stability and defective excitation–contraction coupling (20, 21). From a descriptive/histological point of view, sarcopenia induces a change in the proportion of skeletal muscle fibers, inducing a shift from type II (fast) to type I (slow) fibers as well as preferential loss of type II fibers (22). This fiber-type change impacts not only the strength but also the power of particular muscles, and may be upstream of the signs of mobility limitation observed in sarcopenic patients (23).

12.3.1 EXTRINSIC FACTORS: AGING AS A MULTISYSTEM PROCESS

The aging process affects the physiology of many systems that work together to achieve the main function of the musculoskeletal system; namely, motion. Thus, age-related changes in these systems, especially those that affect neurological, inflammatory, and hormonal behavior may be directly related to sarcopenia.

12.3.1.1 ANDROGEN LEVELS AS EXTRINSIC FACTOR FOR SARCOPENIA

In addition to a natural decrease in testosterone levels due to age, abnormal levels of plasma testosterone are observed in men suffering from late-onset hypogonadism, loss of testicular mass, and endocrine diseases involving low androgen production, accelerated testosterone metabolism or malfunctioning androgen receptor (24, 25). Currently it is recognized that low plasma testosterone is associated with metabolic syndrome, diabetes, and cardiovascular diseases (26); however, the impact of low plasma testosterone levels on skeletal muscle in elderly is still controversial. Because testosterone is the main physiological anabolic hormone, a decline in its plasma concentrations by age must be considered one of the causes for loss of muscle mass and an extrinsic factor for sarcopenia.

Figure 1 shows the relationship between androgen levels and the development of sarcopenia. Androgen supplementation has been observed to exert anabolic actions that enhance muscle strength and increase muscle size clinically (27–31). Although the biological mechanisms underlying androgen action in skeletal muscle remain poorly understood, muscle mass has been observed to be regulated by the normal balance between synthesis and degradation of muscle proteins, a mechanism that is regulated by various systemic hormones. Levels of circulating testosterone, estrogens, IGF-1, vitamin D, and thyroid hormones have been observed to decrease with age (5, 32). While the association between testosterone and increase in skeletal muscle mass has been recognized for many years, it was not formally described until 1996, when Bhasin et al. demonstrated the anabolic effects of testosterone in a study of testosterone-treated vs. placebo-treated young men (9). While this study and subsequent studies provided evidence for anabolic properties of testosterone at all ages, none directly identified the mechanism underlying the relationship between testosterone levels and muscle mass with aging. In a cross-sectional study of 1,445 men, Krasnoff et al. (33) found no significant associations between either total testosterone or SHBG level and mobility, subjective health, or any physical performance measure, but did find an inverse association between free testosterone level and subjective health and a direct association between free testosterone level and usual walking speed and short physical performance battery score. In a subsequent longitudinal analysis, the authors found that free testosterone levels were also significantly associated with progression of mobility limitation (33). Specifically, they found that men with low free testosterone levels were 57% more likely to develop mobility limitation and 68% to experience increase in mobility limitation compared to men with normal testosterone levels. In support of this finding, Malmström et al. found that when the value is adjusted for height, bioavailable testosterone is responsible for 2.6% of the variance in appendicular skeletal muscle mass (ASM), which represents 75% of the total skeletal muscle mass (34).

Clinically, androgen supplementation has been observed to exert anabolic actions that enhance muscle strength and increase muscle size. While elevated plasma androgen concentrations are known to induce skeletal muscle hypertrophy (35), the cellular mechanisms by which testosterone increases skeletal muscle mass remain under investigation. Research to date indicates that testosterone stimulates protein synthesis by both a short-term mechanism-rapid activation of pre-existing components of the translational apparatus- and a long-term mechanism-increase in cell or tissue capacity at the protein synthesis level leading to increase in ribosome quantity (36). At the cellular level, Sinha-Hikim et al. observed that testosterone induces an increase in cross-sectional area (CSA) in type I and II muscle fibers and in myonuclear quantity, indicating that testosterone exerts more of a hypertrophic than a hyperplasic effect on skeletal muscle (30). At mechanistic level, have been observed that testosterone exerts a rapid non-genomic effect on skeletal muscle cells similar to that exerted by other steroid hormones such as estrogen, progesterone, vitamin D3, and aldosterone in different cellular types (37). In rat myotubes, a type of "immature" muscle fiber, our group recently observed rapid phosphorylation of Akt via phosphatidylinositol-3 kinase (PI3K) and subsequent phosphorylation of mammalian target of rapamycin (mTOR), which lies upstream of the critical translation regulator 40S ribosomal protein S6 kinase 1 (S6K1), to increase protein synthesis (38). mTOR represents a key controller of anabolic processes, particularly translation initiation and elongation that produce muscular protein. By activating mTOR, testosterone may induce phosphorylation of eIF4E-binding protein 1 (4E-BP1) and its release from eIF4E, a cap-binding factor that can be sequestered in inactive complexes by 4E-BP1, allowing for generation of initiation factor complexes (39). Moreover, testosterone-induced activation of the PI3K/Akt/mTOR/S6K1 axis combined with "classic" androgen-mediated action in rat myotubes has been observed to lead to hypertrophy, as evidenced by an increase in the CSA of testosterone-treated myotubes and levels of α-actin contractile protein (38). Other pathways have been hypothesized to underlie testosterone-induced hypertrophy. In a study of the cellular line L6, Wu et al. observed that testosterone treatment induced activation of mTOR, but more slowly (within 2 h) than had previous studies and in a manner not mediated by PI3K/Akt but rather MAPK (40). After confirming the anabolic potential of testosterone both in vivo and in vitro, White et al. described the activation of mTOR as dependent on the concentration of testosterone (41). For us, this point is critical to make comparisons between different studies and the effectiveness of therapy testosterone replacement, mainly because differences between the physiological effects of testosterone and changes in its plasma concentration, either low or high levels, may trigger different intracellular mechanisms for testosterone actions on skeletal muscle.

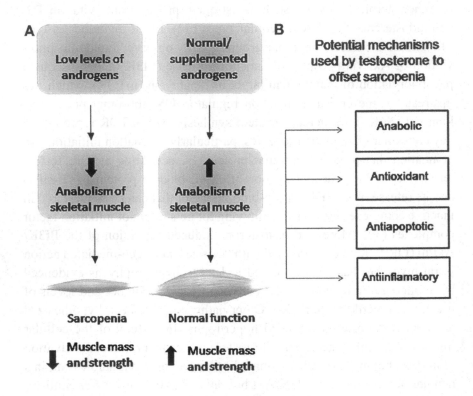

FIGURE 1: (A) Relationship between levels of androgens and anabolism of skeletal muscle and their effects on muscle mass and strength. (B) Cellular effects of testosterone on skeletal muscle.

In summary, recent studies have found that testosterone activates multiple signal transduction pathways to induce hypertrophy, including the MEK-ERK and PI3K/Akt pathways; that the energy sensors mTOR and AMP-activated protein kinase (AMPK) control cell growth and metabolism in testosterone-induced skeletal and cardiac hypertrophy (38, 42); and that alterations in mTOR pathway in aging muscle decrease the kinase activity of mTOR and its downstream targets, decreasing anabolic activity in response to external stimuli, such as growth factor production, nutrient ingestion, and physical activity (43). In the emergent research into the early signals that mediate the activation/deactivation of these essential metabolic pathways, mTOR has been identified as a key element in the regulation of translation and cell growth; namely, as a "master regulator" in a network that detects the energy and nutritional status and genotoxic stress level of cells (44). Identification of the central role of testosterone in metabolism and its effects on mTOR pathway have led it to become a subject of intense interest in aging research and pharmacological treatment for sarcopenia.

12.3.1.2 INFLAMMATORY CYTOKINE PRODUCTION

Loss of skeletal muscle in sarcopenia may result from reduction in muscle protein synthesis and/or increased muscle protein breakdown, both of which are associated with inflammation. In this sense, other hypotheses regarding sarcopenia pathophysiology focus on the role of chronic low-grade inflammation, specifically that related to circulating levels of IL-6 and tumor necrosis factor-α (TNF-α). In a study of the relationship between muscle mass and strength in healthy individuals aged 70–79 years, Visser et al. found IL-6 levels of >1.80 pg/ml and TNF-α levels >3.20 pg/ml to be associated with smaller muscle area, lower appendicular muscle mass, and less strength, even after correcting for presence of inflammatory-associated diseases, height, and smoking (45). One mechanism hypothesized to underlie muscular atrophy in response to elevated TNF-α level is an increase in the ubiquitin-conjugating activity of the E2 protein UbcH2/E220K. In a study using cultured myotubes, Li et al. detected NF-κB mediated up-regulation of UbcH2 at the mRNA level after 90 min of

TNF-α stimulation and at the protein level after 4 h of stimulation, as well as an increase in ubiquitin-conjugating activity after 4–6 h of stimulation to a level that remained elevated for at least 24 h (46). According to the authors, these findings could explain the mechanism whereby TNF-α may induce the loss of skeletal muscle proteins. In line with these findings, in a 5-year follow up study of older healthy individuals Schaap et al. observed a relationship between inflammatory marker (IL-6, TNF-α, and C-reactive protein) levels and a decline in muscle mass and strength, as well as consistent inverse association between levels of TNF-α and its soluble receptors and thigh muscle area and grip strength (47).

12.3.2 INTRINSIC FACTORS: CHANGES AT THE SKELETAL MUSCLE LEVEL

Intrinsic factors leading to alterations in skeletal muscle involve various intracellular signaling pathways, which in turn may activate intracellular modulators in a manner that has long-term cellular effects. The following sections describe several signaling pathways that may be involved in loss of strength due to atrophy of skeletal muscle fibers, a criterion for diagnosis of sarcopenia.

12.3.2.1 MITOCHONDRIAL MALFUNCTION AND OXIDATIVE STRESS

Among the mechanisms hypothesized to underlie atrophic changes in sarcopenic muscles, mechanisms at the mitochondrial level have earned strong research support. In a cross-sectional study comparing active and sedentary elderly adults Safdar et al. (48) found that a sedentary lifestyle was associated with increased basal levels of markers of chronic inflammation (interleukin-6 and C-reactive protein levels), reduced level of markers of mitochondrial oxidative capacity (COX activity and levels of citrate synthase and PGC1-α levels), and deregulation of cellular redox status as superoxide dismutase (SOD) activity. All these changes, which are thought to render skeletal muscle fiber more prone to reactive oxygen species (ROS)-mediated toxicity and, consequently, to skeletal muscle fiber death, led the authors to suggest that mitochondria alterations and increased oxidative status in skeletal muscle are involved in sarcopenia development (48).

Oxidative damage and free radical production have also been associated with a decline in mitochondrial DNA content, protein synthesis, and activity, as well as in oxidative capacity and ATP production (49). In a comparison of aged (20 months-old) $Sod1^{-/-}$ mice, a transgenic mouse model lacking the antioxidant enzyme CuZnSOD, and wild-type mice, Jang et al. observed greater ROS release (O_2^- and H_2O_2) and oxidative damage and decreased mitochondrial bioenergetic functioning and ATP production in the $Sod1^{-/-}$ mice (50). The authors hypothesized that the consequences of the mitochondrial dysfunction that they observed in the $Sod1^{-/-}$ mice, which included elevated apoptotic potential; decrease in myonuclear number per millimeter of fiber length; and alterations in neuromuscular junction histology and contractility, also occur in the skeletal muscle of aging humans, and thus that mitochondrial dysfunction is an important factor in the pathophysiology of sarcopenia.

Recent in vivo and in vitro studies have highlighted the role of testosterone in preventing or reversing age-related skeletal muscle loss. In a recent animal study, a regimen of testosterone administration and simultaneous low-intensity physical training was found to improve skeletal muscle mitochondrial biogenesis and mitochondrial quality control in elderly male mice, suggesting the importance of maintaining proper testosterone levels for muscle metabolism (51). In various cellular models, mainly of prostate cancer cells, low levels of testosterone were found to be associated with oxidative stress induced by decreased antioxidant levels via decreased expression of antioxidant enzymes such as SOD, GSH-Px, and catalase (52). Whereas, testosterone deficiency has been observed to induce oxidative stress in cardiomyocytes, testosterone replacement therapy (TRT) has been found capable of suppressing oxidative stress mediated via the androgen receptor-independent pathway (52).

Skeletal muscle performance has been related to redox status. In soleus muscle of sedentary volunteers, Delgado et al. observed that administration of stanozolol, an anabolic steroid derivate, increased SOD, and glutathione reductase activities, but induced no change in other enzyme activities as catalase and glutathione peroxidase (53). A recent study shows that administration of dihydrotestosterone (DHT) to SOD1-G93A-mutated mice ameliorated muscle atrophy and increased body weight (54). As mutations in the human SOD1 gene are responsible for the development of amyotrophic lateral sclerosis (ALS), a late-onset neurodegenerative disease characterized by progressive loss of motoneurons, skeletal muscle weakness, atrophy, and paralysis, the authors suggested that androgen therapy may improve the quality of life for ALS patients. However, further study of antioxidant regulation by androgens in elderly humans with loss of skeletal muscle is required before extrapolation of these findings to sarcopenic patients.

12.3.2.2 SKELETAL MUSCLE APOPTOSIS

Apoptosis, the process of programed cellular self-destruction without inflammation or damage to surrounding tissue, is highly related to the atrophic process that leads to sarcopenia (55, 56). Despite observation that this cellular response is driven by specific signaling pathways and ultimately executed through caspase-dependent and caspase-independent mechanisms (22), few studies have attempted to confirm the existence of these mechanisms in humans (57). Among the few that have, a cross-sectional study that compared apoptosis levels in the skeletal muscles of elderly and young adults observed a higher number of terminal deoxy-nucleotidyl transferase dUTP nick end labeling (TUNEL)-positive cells and decreased muscle strength in elderly but no significant differences in caspase-dependent pathway-related (caspase-3/7) activity or muscle fiber CSA between the elderly and young (56). In a study of the relationship between apoptotic activation in skeletal muscle and indices of muscle function and mass in the elderly Marzetti et al. found a correlation between biologic apoptosis markers and muscle mass and function indices in community-dwelling elderly men and women (58). Specifically, they found that caspase-dependent apoptotic signaling (caspase-8, cleaved caspase-3, cytosolic cytochrome c, and mitochondrial Bak activity) was correlated with percentage of muscle volume, whereas cleaved caspase-3, Bax, and Bak activity was correlated with gait speed. Although they found no significant differences regarding these apoptotic markers in subjects classified as either high functioning or low functioning according to whether their percentage of muscle volume was below or above the median value of the study sample, they observed higher levels of activated caspase-8 and cytosolic cytochrome c in the low-functioning subjects (57). These findings, in addition to evidence obtained from animal research, support the role of apoptosis as an important mechanism in the pathophysiology of sarcopenia.

Research has also indicated that testosterone has an anti-apoptotic effect in C2C12 muscle cells. Pronsato et al. found that treatment with testosterone at physiological concentrations inhibited apoptosis that had been induced by treatment with a high concentration of H_2O_2 (1 mM) (59). It also induced inactivation of BAD, inhibition of PARP cleavage, decrease in BAX levels, and exertion of a protective effect at the mitochondrial membrane potential level. In this line of evidence, testosterone administration prevented sarcopenia in a mouse model, which allow to hypothesize that testosterone treatment improves the regenerative potential of satellite cells and suppresses skeletal muscle cell apoptosis, the latter primarily by inhibiting activation of the JNK pathway and activating Akt (60). Further research has confirmed that testosterone supplementation exerts both beneficial and pathological effects on apoptotic process. Depletion of androgen below normal plasma levels has been found to increase kainate-induced neuronal loss (61) and induce apoptosis in the bladder wall of senile male rats (62), while elevated testosterone has been observed to induce apoptosis in a human neuroblastoma cell line (63). These findings indicate that the beneficial effects of androgen are obtained when androgen levels are within a specific (nanomolar) physiological range, the regulation of which is crucial for normal muscular functioning and a deviation from which in either direction could be deleterious to muscle health.

12.3.2.3 MUSCLE ATROPHY AND CA²⁺ SIGNALING

The cellular processes underlying muscle atrophy result in both qualitative and quantitative alterations in muscle fiber cells and associated structures. Muscle atrophy leaves particular "signatures" in the profile of sarcopenic muscle proteins that can be observed, and muscle proteins are contained in different cellular compartments that can be analyzed. Since contractile proteins are critical for skeletal muscle function, analyzing changes in these proteins can assist in understanding the pathophysiology of sarcopenia. In a study of the synthesis of myosin heavy chain (MHC) protein over the lifespan and its correlation with muscle strength and hormone levels, Balagopal et al. (64) found that the rate of MHC synthesis begins decreasing at age 50 before plateauing at age 75, and observed a correlation between a decrease in the rate of MHC synthesis and muscle mass, strength measures and, interestingly, levels of anabolic hormones, including insulin-like growth factor I (IGF-I) and dehydroepiandrosterone sulfate (DHEAS) levels in both men and women and free testosterone level in men. As myosin represents about 25–30% of muscle protein, a decrease in its synthesis rate would be expected to impact muscle mass and strength (64). Other protein-specific changes in senescent skeletal muscle related to decreased muscle function involve α-actin, Ca-ATPase transporter, ryanodine receptor, and muscle-specific inositol phosphatase (MIP), a recently described protein related to Ca^{2+} homeostasis in skeletal muscle. In a recent study, Staunton et al. observed a lower level of thin filament α-actin protein in vastus lateralis muscle biopsy samples taken from aged subjects compared to those taken from middle-aged subjects (65). Utilizing mass spectrometry-based proteomic analysis, the authors were able to describe a "signature" of aged vastus lateralis skeletal muscle and detect changes in proteins involved in excitation–contraction coupling, metabolism, ion handling, and the cellular stress response. These findings led the authors to suggest that human muscle tissue undergoes a fiber-type shift from fast to slow contracting muscle tissue with age, an event that has been previously postulated and demonstrated in other sarcopenic models (65).

Concerning intracellular Ca^{2+} handling proteins in skeletal muscle, Ferrington et al. (66) described a decrease in the relative protein turnover of the sarcoplasmic reticulum proteins Ca-ATPase and the ryanodine receptor in aged rats. The authors hypothesized that alterations in the turnover rate of certain key Ca^{2+} handling proteins could account for the functional disturbances observed in skeletal muscle. Our research group has demonstrated that androgens are able to modulate intracellular Ca^{2+} homeostasis within seconds to minutes in different cell systems using a variety of mechanisms that vary considerably and depend on the cell type (67–69). In skeletal muscle, we observed that testosterone stimulation induces rapid (<1 min) oscillation in intracellular Ca^{2+} signals, which begin as Ca^{2+} transients initiated in the cytosol before being propagated as waves of Ca^{2+} in the cytoplasm and nucleus (68, 70, 71). This complex process of Ca^{2+} signaling, which depends on the interplay between IP3-sensitive stores and extracellular Ca^{2+} influx, is unaffected by the androgen-receptor antagonist cyproterone acetate, and apparently mediated by a pertussis toxin-sensitive G protein-coupled membrane receptor activated by testosterone 3-(O-carboxymethyl)oxime (T-BSA). Moreover, it has been suggested that a similar manner of conformational coupling between IP3 receptors and store-operated calcium entry (SOCE) is activated in skeletal myotubes during testosterone-induced Ca2+ oscillations (70). In this sense, age-related testosterone decrease certainly will affect skeletal muscle Ca^{2+} homeostasis.

As described by Shen (72) (MIP/MTMR14), a recently described protein, is responsible for sarcopenia pathophysiology by controlling intracellular phosphatidylinositol phosphate (PIP) levels via influencing SOCE and Ca^{2+} storage and release from the sarcoplasmic reticulum (72). Its importance is evidenced by observation of significant age-related decreases in levels of MIP mRNA, protein content, and activity in wild-type mice. Characteristic features of sarcopenia (diminished muscle mass, force, and power generation) also appear much earlier in MIP knock-out (MIPKO) mice than in their wild-type counterparts. Altered Ca^{2+} homeostasis has also been observed in both mature MIPKO and old wild-type mice (21). Moreover, muscle aging has been associated with compromised Ca^{2+} spark signaling and segregated intracellular Ca^{2+} release in sarcopenia (73). According to Romero-Suarez et al. (21), these findings all provide evidence of the important role of MIP in the physiology and pathology of muscle fiber.

12.3.2.4 AGE-RELATED DEPLETION OF MUSCLE STEM CELLS

The sarcopenic process is more pronounced in patients showing significant loss of muscle stem cells, particularly satellite cells. Several reports indicate that it is possible to isolate stem cells from adult skeletal muscle, but is important to distinguish between satellite cells and muscle-derived stem cells (MDSC). Embryonic stem cells and induced pluripotent stem cells are another cell source that can differentiate into muscle cells, but further work is needed to produce sufficient numbers of cells to generate contractile myofibers for therapeutic use (74, 75).

The satellite cells are mononucleated cells that lie under or embedded in the basal lamina of the myofiber, which demonstrate close relationship with the mature myofiber (75). The satellite cells are myogenic precursors capable of regenerating skeletal muscle and demonstrate self-renewal properties. During development and regeneration, quiescent satellite cells are activated and start proliferating, at this point they are called myogenic precursor cells or myoblasts (76). Once muscle satellite cells are activated to become myoblasts, they enter the proliferative stage and differentiate into myotubes by expression of MyoD, whereas, the secondary myogenic regulatory factors (MRF) as myogenin and MRF4 regulate terminal differentiations. Myofibers derived from satellite cells show characteristic skeletal muscle markers such as sarcomeric striations, MHC, MyoD, and desmin expression (77).

In the elderly, the number of satellite cells and their capacity of cellular regeneration decrease (78). In humans and mice, these quiescent cells are plentiful at birth but their number declines with age until 1–5% in adults (74). Different modulators regulate satellite cell functions in adults. Among them Alway et al. suggest that satellite cell function is affected by oxidative stress, which is elevated in aged muscles (79). During the differentiation of satellite cells to muscle a normal mitochondrial oxidative metabolism, with low production of ROS, is required to sustain skeletal muscle specification and function. However, aging is associated with excessive production of ROS by increase in mitochondrial damage contributing to impaired satellite cell function (79). Also, the decrease in satellite cells by age, has been attributed to the change in percentage of type I vs. type II muscle fibers, because satellite cells reside surrounding type II muscle fibers (46, 78).

Characterization of this satellite cells-derived skeletal muscle is determined at molecular, electrophysiological, and functional levels. At the molecular level, they show muscle-specific markers including transcription factors such as Pax7, Pax3, c-Met, M-cadherin, CD34, Syndecan-3, and calcitonin (80, 81). Recent experiments showed that, in contrast to cultured myoblasts, satellite cells freshly isolated or satellite cells derived from the transplantation of one intact myofibre, contribute robustly to muscle repair. However, because satellite cells are known to be heterogeneous, clonal analysis is required to demonstrate stem cell function.

There is consensus that the use of testosterone leads ultimately to regulate skeletal muscle mass by a net increase in protein synthesis over degradation. In addition for changes at the skeletal muscle level, it has been described that other cell types are involved in testosterone-induced muscle functions. Testosterone has direct effects on satellite cells, because they express the androgen receptor and in response to testosterone increase the satellite cell population. This cell proliferation is followed by a subsequent increase in the myonuclei number of the mature skeletal muscle, through the fusion of the satellite cells with pre-existing fibers resulting in muscle hypertrophy (12, 54, 82). Mesenchymal stem cells (MSCs) are a group of non-hematopoietic stem cells residing in bone marrow that can be used to treat a variety of degenerative diseases, including musculoskeletal diseases. The mechanism proposed to explain the effect of testosterone on fat-free mass, highlights the involvement of the androgen receptor in the commitment of an undifferentiated cell type into a myogenic cell line (83, 84).

12.4 THERAPEUTIC APPROACHES IN SARCOPENIA

Age-related frailty and sarcopenia have emerged as important public health concerns because of their negative impact on mobility, quality of life, and health care resources (85, 86). Nutrition and physical activity are two modifiable factors that can affect the development of sarcopenia. Improving either or both, nutrition and physical activity, may reduce the age-related low-grade chronic inflammation and/or activate the intrinsic anabolic pathways in skeletal muscle. While resistance exercise, which has assumed a prominent role in the treatment of sarcopenia, has been shown to improve skeletal muscle mass and strength in sarcopenic patients, it has not always been observed to improve functional parameters (87).

Androgen treatment has been observed to enhance skeletal muscle strength and size (30, 88, 89), but the biological mechanisms underlying androgen action in skeletal muscle are not completely known. In a review of six studies of TRT in elderly men, Borst reported that several studies found that testosterone treatment increased lean mass and decreased fat mass whereas one study observed increased strength (7).

12.4.1 ANABOLIC INTERVENTIONS

Muscle plasticity and regeneration are key processes in a number of myopathies, including muscular dystrophy, neuromuscular disorders, and aging. Recent years have witnessed a growing interest in using anabolic interventions to counteract loss of muscle mass and function associated with age-related decreases in testosterone levels in men. Observation that low levels of testosterone are associated with decreases in fat-free mass, ASM, and strength in hypogonadal males compared to healthy controls has served as the basis to use TRT to treat hypogonadal men (90). With age, not only do testosterone levels decline progressively but SHBG levels also increase, further decreasing the amount of bioavailable testosterone. The prevalence of hypogonadism is approximately 20% in men over 60 years and can reach 50% in men over 80 years (31). In young hypogonadal men, TRT has been associated with increase in lean mass, muscle strength, and muscle protein synthesis and decrease in fat mass. However, in eugonadal men, changes in body composition following TRT have not always been followed by increase in muscle strength (91), leading to controversy regarding the ergogenic effect of TRT. In one study, TRT increased bone mass only in the group of patients with hypogonadism (89). In addition, several studies found that administration of supraphysiological doses of testosterone to hypogonadal patients resulted in outcomes similar to those obtained with resistance exercise (9).

Testosterone replacement therapy can be administered through several routes, including intramuscularly, transdermally, and orally (i.e., via testosterone undecanoate). Unfortunately, few studies of the safety of TRT using any of these routes, specifically its association with prostate cancer and cardiovascular disease in the elderly have been conducted. Moreover,

the few studies that have been conducted were only observational or examined administration of only very high levels of testosterone. To date,
TRT has been observed to induce and exacerbate sleep apnea, transient
fluid retention, gynecomastia, increase red cell mass, and increase the size
of both benign and malignant prostate tumors (31). TRT for hypogonadal
patients has also been hypothesized to increase cardiovascular risk through
its effect on lipid metabolism.

Many studies reporting an increase in muscle strength with TRT suf-
fered from methodological problems, such as lack of a control group, lack
of control for the effects of exercise, lack of control of the dose(s) of the
hormones administered to maintain normal levels of circulating testosterone, or the inclusion of a very small number of patients. In a meta-analysis
that pooled data from 19 randomized placebo-controlled trials, Calof et al.
(28) reported that compared to placebo-treated men, testosterone-treated
men were found 1.8 times more likely to experience prostate events, including prostate cancer, elevated prostate-specific antigen level, and prostate biopsy, but not at a level that differed significantly at the individual
level. They also found that testosterone-treated men were, 3.67 times more
likely to experience hematocrit over 50%, the most common testosterone-
related adverse event, but did not experience cardiovascular events at a
significantly higher rate (28).

In a randomized placebo-controlled trial of 44 men with late-onset hypogonadism aged 44–78 years, Marks et al. (92) found that 6 months of
TRT normalized serum testosterone levels while only slightly and insigni-
ficantly increasing androgen levels in prostate tissue and volume, prostate-
specific antigen, voiding symptoms, and affecting urinary flow, and did
not alter prostate tissue composition, biomarkers of cell proliferation, gene
expression, or cancer incidence or severity (92, 93).

Observation that levels of dehydroepiandrosterone (DHEA), a precursor of various sex steroids produced in the adrenal cortex, decline gradually with age from the third decade of life has motivated research examining whether DHEA supplementation can reverse the pathophysiological
changes associated with aging. It is hypothesized that DHEA supplementation can increase muscle strength by increasing ratio of circulating testosterone to cortisol. In a study of men and women aged 50–65 years,
(90) observed that supplementation with 100 mg of DHEA for 6 months

increased lean body mass and decreased fat mass in both sexes, but increased muscle strength only moderately and only in men, and increased testosterone levels only in women. In a randomized placebo-controlled trial of 50 mg/day of DHEA supplementation for 1 year to men and women aged 60–80 years, (94) failed to reproduce Morales's results or detect an increase in lean mass, as indicated by measurements of body potassium.

Oxandrolone, an androgenic steroid with powerful anabolic effects that can be administered orally, is resistant to hepatic metabolism and thus less hepatotoxic compared to other oral androgens. Moreover, its side effects, which include discrete elevations in transaminase levels and decreased high-density lipoprotein cholesterol levels, are mild and transient (95). Despite these advantages, no clinical studies of oxandrolone administration to elderly patients with sarcopenia have been conducted to date. However, previous studies of skeletal muscle wasting in cachexia associated with HIV infection, chronic neuromuscular conditions, and disease-related loss of muscle mass have indicated that oxandrolone increases protein synthesis in skeletal muscle, and is thus associated with increased physical activity capacity and level, increased energy and protein intake, reduced visceral and total fat mass, and improvement in nitrogen retention (96). Therefore, oxandrolone administration may be a therapeutic strategy in the treatment of sarcopenia in the elderly.

Supplementation with selective androgen-receptor modulators (SARMs) has emerged as a means of treating muscle and bone disorders, mainly because of the specificity of SARM action and the relatively few side effects of SARM treatment. Research using experimental models has demonstrated that administration of SARM S-4 exerts potent anabolic effects on skeletal muscle and bone and only minimal effects on the prostate (97). In a recent randomized, double-blind, placebo-controlled study of 170 women aged ≥65 with sarcopenia and moderate physical dysfunction found that 6 months of treatment with MK-0773 significantly improved physical performance measures (98). GTx-024 (enobosarm), a non-steroidal SARM that exerts tissue-selective anabolic effects in muscle and bone while sparing other androgenic tissue related to hair growth in women and prostate effects in men, has demonstrated promising pharmacologic effects in preclinical studies and favorable safety and pharmacokinetic profiles in phase I investigations. Thus GTx-024 supplementation resulted in dose-dependent improvement in total lean body mass and physical function and was well tolerated (99). Treatment with one or more of the numerous other SARMs currently under study may emerge as therapeutic alternatives to androgen agonist therapy. The intense research into pharmacological modulation of androgens and androgen intracellular signaling pathways may lead to the development of effective approaches to restoring and preventing the muscle loss observed in sarcopenia.

12.4.2 TREATMENT ALTERNATIVES AND CONCERNS

Because sarcopenia is a multifactorial disease, this should be treated using a variety of therapeutic approaches including diet, exercise, and pharmacology. Among current therapies, resistance exercise and androgen replacement therapy appear to be effective alternatives for sarcopenia because both modulate multiple tissues including musculoskeletal function. Therefore, in older men with sarcopenia a program of physical exercise together with TRT emerges as a promising therapeutic alternative. A great concern over the previous decade was that androgen supplementation increased risk of cardiovascular and prostate events. However, recent research has shown no association between androgen supplementation and increased cardiovascular risk (with several studies even identifying an association between low levels of testosterone and cardiovascular risk) nor between androgen supplementation and prostate disease. The development of "state-of-the-art" testosterone treatments has called for wide-scale longitudinal population studies to determine their profile. While other anabolic, such as DHEA, oxandrolone, and SARMs appear to be promising agents in sarcopenia treatment, further research is required before recommendations regarding their pharmacological use can be developed.

A decline in homeostatic and regenerative capacity occurs in aging, where a degenerative change in stem cells homeostasis has been postulated. Cell replacement therapy is a potential alternative treatment currently undergoing clinical trials (100). However, in elderly human muscle they are in extremely limited supply, hence there is a high demand for an alternative satellite cell source. There are a number of potential sources of muscle stem cells for cell replacement therapies such as bone marrow-derived stem cells, hematopoietic stem cells, and MSCs. However, there are significant controversies regarding both the efficiency and the reality of skeletal muscle differentiation by many of these stem cell types. Among these cell sources, satellite cells unquestionably undergo the most efficient musculogenesis. The actual focus in stem cell therapy, implicate to enhance satellite cell activity by environmental conditions and stem cell transplant into damaged tissues. Recently, it has been described that satellite cells can be transplanted into the muscle of mice, and they are able to proliferate. Moreover, these stem cells generate a massive proliferation during muscle injury (100). A recent in vitro study showed that treatment of 10T1/2 pluripotent mesenchymal cells with testosterone or DHT significantly increased the number of myogenic cells in a dose-dependent manner, while inhibiting adipogenic differentiation (101, 102). This study demonstrated that these effects are mediated through an AR-dependent mechanism, because an AR antagonist blocked the actions of testosterone or DHT. This hypothesis is also supported by the fact that in humans, CD34+ interstitial, mesenchymal cells are AR positive and expression of the AR is androgen dose-dependent (13, 102).

These data demonstrate that androgens can recruit stem cells into the myogenic lineage by committing them to myogenic precursor cells (101, 103). However, it is still not known whether androgen stimulation of myogenic commitment then gives rise to satellite cells, or directly contributes to muscle formation. Moreover, the physiological significance of the effects of androgens on stem cell commitment in contributing to muscle growth and regeneration is unclear. These preliminary studies clearly indicate that further investigation into this area is promissory and necessary.

12.5 CONCLUSION

As an emerging syndrome with an increasing prevalence, sarcopenia has the potential to affect all aging adults. Therefore, we must consider that in older men with low testosterone levels and symptoms of androgen deficiency, hormone replacement therapy in combination with physical activity and proper nutrition will result in increased muscle mass and strength. However, data on the clinical effects of androgen replacement therapy to physiological ranges are not yet available. With the increasing aging of most of the world's populations, research into this disabling disease, which not only decreases quality of life but also increases risk of mortality, is urgently required. Identification of the pathophysiological mechanisms underlying sarcopenia and the development of therapeutic approaches will improve the quality of life not only of the current elderly population, but all of us when we walked into our "golden years".

12.5.1 HIGHLIGHTS

- Sarcopenia is a decrease in skeletal muscle mass and function associated with age.
- Local and systemic factors as testosterone levels are associated with sarcopenia.
- No one of the several treatments for sarcopenia has been found clearly superior.
- Exercise, proper nutrition, and testosterone supplementation may treat sarcopenia.

REFERENCES

1. Lexell J, Henriksson-Larsen K, Winblad B, Sjostrom M. Distribution of different fiber types in human skeletal muscles: effects of aging studied in whole muscle cross sections. Muscle Nerve (1983) 6:588–95. doi:10.1002/mus.880060809
2. Rosenberg IH. Sarcopenia: origins and clinical relevance. J Nutr (1997) 127:990S–1S.
3. Rosenberg IH. Summary comments. Am J Clin Nutr (1989) 50(Suppl):1231S–3S.
4. Baumgartner RN, Koehler KM, Gallagher D, Romero L, Heymsfield SB, Ross RR, et al. Epidemiology of sarcopenia among the elderly in New Mexico. Am J Epidemiol (1998) 147:755–63. doi:10.1093/oxfordjournals.aje.a009520
5. Degens H, Korhonen MT. Factors contributing to the variability in muscle ageing. Maturitas (2012) 73:197–201. doi:10.1016/j.maturitas.2012.07.015
6. Samaras N, Papadopoulou MA, Samaras D, Ongaro F. Off-label use of hormones as an antiaging strategy: a review. Clin Interv Aging (2014) 9:1175–86. doi:10.2147/CIA.S48918
7. Borst SE. Interventions for sarcopenia and muscle weakness in older people. Age Ageing (2004) 33:548–55. doi:10.1093/ageing/afh201
8. Lippi G, Sanchis-Gomar F, Montagnana M. Biological markers in older people at risk of mobility limitations. Curr Pharm Des (2014) 20:3222–44. doi:10.2174/1381 6128113196660697

9. Bhasin S, Storer TW, Berman N, Callegari C, Clevenger B, Phillips J, et al. The effects of supraphysiologic doses of testosterone on muscle size and strength in normal men. N Engl J Med (1996) 335:1–7. doi:10.1056/NEJM199607043350101

10. Sinclair M, Grossmann M, Gow PJ, Angus PW. Testosterone in men with advanced liver disease: abnormalities and implications. J Gastroenterol Hepatol (2014). doi:10.1111/jgh.12695

11. Urban RJ, Dillon EL, Choudhary S, Zhao Y, Horstman AM, Tilton RG, et al. Translational studies in older men using testosterone to treat sarcopenia. Trans Am Clin Climatol Assoc (2014) 125:27–44. Available from: http://www.ncbi.nlm.nih.gov/pmc/articles/PMC4112698/

12. Yuki A, Ando F, Otsuka R, Shimokata H. Low free testosterone is associated with loss of appendicular muscle mass in Japanese community-dwelling women. Geriatr Gerontol Int (2014). doi:10.1111/ggi.12278

13. Sinha I, Sinha-Hikim AP, Wagers AJ, Sinha-Hikim I. Testosterone is essential for skeletal muscle growth in aged mice in a heterochronic parabiosis model. Cell Tissue Res (2014) 357(3):815–21. doi:10.1007/s00441-014-1900-2

14. Cruz-Jentoft AJ, Baeyens JP, Bauer JM, Boirie Y, Cederholm T, Landi F, et al. Sarcopenia: European consensus on definition and diagnosis: report of the European working group on sarcopenia in older people. Age Ageing (2010) 39:412–23. doi:10.1093/ageing/afq034

15. Atkinson RA, Srinivas-Shankar U, Roberts SA, Connolly MJ, Adams JE, Oldham JA, et al. Effects of testosterone on skeletal muscle architecture in intermediate-frail and frail elderly men. J Gerontol A Biol Sci Med Sci (2010) 65:1215–9. doi:10.1093/gerona/glq118

16. Janssen I, Heymsfield SB, Wang ZM, Ross R. Skeletal muscle mass and distribution in 468 men and women aged 18–88 yr. J Appl Physiol (1985) (2000) 89:81–8. Available from: http://jap.physiology.org/content/89/1/81

17. Liu LK, Lee WJ, Chen LY, Hwang AC, Lin MH, Peng LN, et al. Sarcopenia, and its association with cardiometabolic and functional characteristics in Taiwan: results from I-Lan Longitudinal Aging Study. Geriatr Gerontol Int (2014) 14(Suppl 1):36–45. doi:10.1111/ggi.12208

18. Janssen I, Baumgartner RN, Ross R, Rosenberg IH, Roubenoff R. Skeletal muscle cutpoints associated with elevated physical disability risk in older men and women. Am J Epidemiol (2004) 159:413–21. doi:10.1093/aje/kwh058

19. Feldman HA, Longcope C, Derby CA, Johannes CB, Araujo AB, Coviello AD, et al. Age trends in the level of serum testosterone and other hormones in middle-aged men: longitudinal results from the Massachusetts male aging study. J Clin Endocrinol Metab (2002) 87:589–98. doi:10.1210/jcem.87.2.8201

20. Delbono O. Molecular mechanisms and therapeutics of the deficit in specific force in ageing skeletal muscle. Biogerontology (2002) 3:265–70. doi:10.1023/A:1020189627325

21. Romero-Suarez S, Shen J, Brotto L, Hall T, Mo C, Valdivia HH, et al. Muscle-specific inositide phosphatase (MIP/MTMR14) is reduced with age and its loss accelerates skeletal muscle aging process by altering calcium homeostasis. Aging (Albany NY) (2010) 2:504–13. Available from: http://www.ncbi.nlm.nih.gov/pmc/articles/PMC2954041/?report=reader#!po=1.25000

22. Marzetti E, Leeuwenburgh C. Skeletal muscle apoptosis, sarcopenia and frailty at old age. Exp Gerontol (2006) 41:1234–8. doi:10.1016/j.exger.2006.08.011

23. Bean JF, Leveille SG, Kiely DK, Bandinelli S, Guralnik JM, Ferrucci L. A comparison of leg power and leg strength within the InCHIANTI study: which influences mobility more? J Gerontol A Biol Sci Med Sci (2003) 58:728–33. doi:10.1093/gerona/58.8.M728

24. Burns-Cox N, Gingell C. The andropause: fact or fiction? Postgrad Med J (1997) 73:553–6. doi:10.1136/pgmj.73.863.553

25. Wu FC, Tajar A, Beynon JM, Pye SR, Silman AJ, Finn JD, et al. Identification of late-onset hypogonadism in middle-aged and elderly men. N Engl J Med (2010) 363:123–35. doi:10.1056/NEJMoa0911101

26. Brand JS, Rovers MM, Yeap BB, Schneider HJ, Tuomainen TP, Haring R, et al. Testosterone, sex hormone-binding globulin and the metabolic syndrome in men: an individual participant data meta-analysis of observational studies. PLoS One (2014) 9:e100409. doi:10.1371/journal.pone.0100409

27. Mauras N, Hayes V, Welch S, Rini A, Helgeson K, Dokler M, et al. Testosterone deficiency in young men: marked alterations in whole body protein kinetics, strength, and adiposity. J Clin Endocrinol Metab (1998) 83:1886–92. doi:10.1210/jcem.83.6.4892

28. Calof OM, Singh AB, Lee ML, Kenny AM, Urban RJ, Tenover JL, et al. Adverse events associated with testosterone replacement in middle-aged and older men: a meta-analysis of randomized, placebo-controlled trials. J Gerontol A Biol Sci Med Sci (2005) 60:1451–7. doi:10.1093/gerona/60.11.1451

29. Nnodim JO. Quantitative study of the effects of denervation and castration on the levator ani muscle of the rat. Anat Rec (1999) 255:324–33.

30. Sinha-Hikim I, Artaza J, Woodhouse L, Gonzalez-Cadavid N, Singh AB, Lee MI, et al. Testosterone-induced increase in muscle size in healthy young men is associated with muscle fiber hypertrophy. Am J Physiol Endocrinol Metab (2002) 283:E154–64. doi:10.1152/ajpendo.00502.2001

31. Surampudi PN, Wang C, Swerdloff R. Hypogonadism in the aging male diagnosis, potential benefits, and risks of testosterone replacement therapy. Int J Endocrinol (2012) 2012:625434. doi:10.1155/2012/625434

32. Marcell TJ, Harman SM, Urban RJ, Metz DD, Rodgers BD, Blackman MR. Comparison of GH, IGF-I, and testosterone with mRNA of receptors and myostatin in skeletal muscle in older men. Am J Physiol Endocrinol Metab (2001) 281:E1159–64. Available from: http://ajpendo.physiology.org/content/281/6/E1159

33. Krasnoff JB, Basaria S, Pencina MJ, Jasuja GK, Vasan RS, Ulloor J, et al. Free testosterone levels are associated with mobility limitation and physical performance in community-dwelling men: the Framingham Offspring Study. J Clin Endocrinol Metab (2010) 95:2790–9. doi:10.1210/jc.2009-2680

34. Malmström TK, Miller DK, Herning MM, Morley JE. Low appendicular skeletal muscle mass (ASM) with limited mobility and poor health outcomes in middle-aged African Americans. J Cachexia Sarcopenia Muscle (2013) 4(3):179–86. doi:10.1007/s13539-013-0106-x

35. Bhasin S, Woodhouse L, Casaburi R, Singh AB, Bhasin D, Berman N, et al. Testosterone dose-response relationships in healthy young men. Am J Physiol Endo-

crinol Metab (2001) 281:E1172–81. Available from: http://ajpendo.physiology.org/content/281/6/E1172.long

36. Isidori AM, Giannetta E, Greco EA, Gianfrilli D, Bonifacio V, Isidori A, et al. Effects of testosterone on body composition, bone metabolism and serum lipid profile in middle-aged men: a meta-analysis. Clin Endocrinol (Oxf) (2005) 63:280–93. doi:10.1111/j.1365-2265.2005.02339.x

37. Wehling M. Specific, nongenomic actions of steroid hormones. Annu Rev Physiol (1997) 59:365–93. doi:10.1146/annurev.physiol.59.1.365

38. Basualto-Alarcon C, Jorquera G, Altamirano F, Jaimovich E, Estrada M. Testosterone signals through mTOR and androgen receptor to induce muscle hypertrophy. Med Sci Sports Exerc (2013) 45(9):1712–20. doi:10.1249/MSS.0b013e31828cf5f3

39. Ma XM, Blenis J. Molecular mechanisms of mTOR-mediated translational control. Nat Rev Mol Cell Biol (2009) 10:307–18. doi:10.1038/nrm2672

40. Wu Y, Bauman WA, Blitzer RD, Cardozo C. Testosterone-induced hypertrophy of L6 myoblasts is dependent upon Erk and mTOR. Biochem Biophys Res Commun (2010) 400:679–83. doi:10.1016/j.bbrc.2010.08.127

41. White JP, Gao S, Puppa MJ, Sato S, Welle SL, Carson JA. Testosterone regulation of Akt/mTORC1/FoxO3a signaling in skeletal muscle. Mol Cell Endocrinol (2013) 365:174–86. doi:10.1016/j.mce.2012.10.019

42. Wilson C, Contreras-Ferrat A, Venegas N, Osorio-Fuentealba C, Pávez M, Montoya K, et al. Testosterone increases GLUT4-dependent glucose uptake in cardiomyocytes. J Cell Physiol (2013) 228(12):2399–407. doi:10.1002/jcp.24413

43. Sandri M, Barberi L, Bijlsma AY, Blaauw B, Dyar KA, Milan G, et al. Signalling pathways regulating muscle mass in ageing skeletal muscle. The role of the IGF1-Akt-mTOR-FoxO pathway. Biogerontology (2013) 14(3):303–23. doi:10.1007/s10522-013-9432-9

44. Weigl LG. Lost in translation: regulation of skeletal muscle protein synthesis. Curr Opin Pharmacol (2012) 12:377–82. doi:10.1016/j.coph.2012.02.017

45. Visser M, Pahor M, Taaffe DR, Goodpaster BH, Simonsick EM, Newman AB, et al. Relationship of interleukin-6 and tumor necrosis factor-alpha with muscle mass and muscle strength in elderly men and women: the Health ABC Study. J Gerontol A Biol Sci Med Sci (2002) 57:M326–32. doi:10.1093/gerona/57.5.M326

46. Li YP, Lecker SH, Chen Y, Waddell ID, Goldberg AL, Reid MB. TNF-alpha increases ubiquitin-conjugating activity in skeletal muscle by up-regulating UbcH2/E220k. FASEB J (2003) 17:1048–57. doi:10.1096/fj.02-0759com

47. Schaap LA, Pluijm SM, Deeg DJ, Harris TB, Kritchevsky SB, Newman AB, et al. Higher inflammatory marker levels in older persons: associations with 5-year change in muscle mass and muscle strength. J Gerontol A Biol Sci Med Sci (2009) 64:1183–9. doi:10.1093/gerona/glp097

48. Safdar A, Hamadeh MJ, Kaczor JJ, Raha S, Debeer J, Tarnopolsky MA. Aberrant mitochondrial homeostasis in the skeletal muscle of sedentary older adults. PLoS One (2010) 5:e10778. doi:10.1371/journal.pone.0010778

49. Drey M. Sarcopenia – pathophysiology and clinical relevance. Wien Med Wochenschr (2011) 161:402–8. doi:10.1007/s10354-011-0002-y

50. Jang YC, Lustgarten MS, Liu Y, Muller FL, Bhattacharya A, Liang H, et al. Increased superoxide in vivo accelerates age-associated muscle atrophy through mi-

tochondrial dysfunction and neuromuscular junction degeneration. FASEB J (2010) 24:1376–90. doi:10.1096/fj.09-146308

51. Guo W, Wong S, Li M, Liang W, Liesa M, Serra C, et al. Testosterone plus low-intensity physical training in late life improves functional performance, skeletal muscle mitochondrial biogenesis, and mitochondrial quality control in male mice. PLoS One (2013) 7:e51180. doi:10.1371/journal.pone.0051180

52. Zhang L, Wu S, Ruan Y, Hong L, Xing X, Lai W. Testosterone suppresses oxidative stress via androgen receptor-independent pathway in murine cardiomyocytes. Mol Med Rep (2011) 4:1183–8. doi:10.3892/mmr.2011.539

53. Delgado J, Saborido A, Megias A. Prolonged treatment with the anabolic-androgenic steroid stanozolol increases antioxidant defences in rat skeletal muscle. J Physiol Biochem (2010) 66:63–71. doi:10.1007/s13105-010-0010-1

54. Yoo YE, Ko CP. Dihydrotestosterone ameliorates degeneration in muscle, axons and motoneurons and improves motor function in amyotrophic lateral sclerosis model mice. PLoS One (2012) 7:e37258. doi:10.1371/journal.pone.0037258

55. Lenk K, Schuler G, Adams V. Skeletal muscle wasting in cachexia and sarcopenia: molecular pathophysiology and impact of exercise training. J Cachexia Sarcopenia Muscle (2010) 1:9–21. doi:10.1007/s13539-010-0007-1

56. Whitman SA, Wacker MJ, Richmond SR, Godard MP. Contributions of the ubiquitin-proteasome pathway and apoptosis to human skeletal muscle wasting with age. Pflugers Arch (2005) 450:437–46. doi:10.1007/s00424-005-1473-8

57. Marzetti E, Privitera G, Simili V, Wohlgemuth SE, Aulisa L, Pahor M, et al. Multiple pathways to the same end: mechanisms of myonuclear apoptosis in sarcopenia of aging. ScientificWorldJournal (2010) 10:340–9. doi:10.1100/tsw.2010.27

58. Marzetti E, Lees HA, Manini TM, Buford TW, Aranda JM Jr, Calvani R, et al. Skeletal muscle apoptotic signaling predicts thigh muscle volume and gait speed in community-dwelling older persons: an exploratory study. PLoS One (2012) 7:e32829. doi:10.1371/journal.pone.0032829

59. Pronsato L, Boland R, Milanesi L. Testosterone exerts antiapoptotic effects against H2O2 in C2C12 skeletal muscle cells through the apoptotic intrinsic pathway. J Endocrinol (2012) 212:371–81. doi:10.1530/JOE-11-0234

60. Kovacheva EL, Hikim AP, Shen R, Sinha I, Sinha-Hikim I. Testosterone supplementation reverses sarcopenia in aging through regulation of myostatin, c-Jun NH2-terminal kinase, Notch, and Akt signaling pathways. Endocrinology (2010) 151:628–38. doi:10.1210/en.2009-1177

61. Ramsden M, Shin TM, Pike CJ. Androgens modulate neuronal vulnerability to kainate lesion. Neuroscience (2003) 122:573–8. doi:10.1016/j.neuroscience.2003.08.048

62. Lorenzetti F, Pintarelli VL, Seraphim DC, Dambros M. Low testosterone levels induce apoptosis via active 3-caspase dependent signaling in the bladder wall of male rats. Aging Male (2012) 15:216–9. doi:10.3109/13685538.2012.716876

63. Estrada M, Varshney A, Ehrlich BE. Elevated testosterone induces apoptosis in neuronal cells. J Biol Chem (2006) 281:25492–501. doi:10.1074/jbc.M603193200

64. Balagopal P, Nair KS, Stirewalt WS. Isolation of myosin heavy chain from small skeletal muscle samples by preparative continuous elution gel electrophoresis: application to measurement of synthesis rate in human and animal tissue. Anal Biochem (1994) 221:72–7.

65. Staunton L, Zweyer M, Swandulla D, Ohlendieck K. Mass spectrometry-based proteomic analysis of middle-aged vs. aged vastus lateralis reveals increased levels of carbonic anhydrase isoform 3 in senescent human skeletal muscle. Int J Mol Med (2012) 30:723–33. doi:10.3892/ijmm.2012.1056

66. Ferrington DA, Krainev AG, Bigelow DJ. Altered turnover of calcium regulatory proteins of the sarcoplasmic reticulum in aged skeletal muscle. J Biol Chem (1998) 273:5885–91. doi:10.1074/jbc.273.10.5885

67. Altamirano F, Oyarce C, Silva P, Toyos M, Wilson C, Lavandero S, et al. Testosterone induces cardiomyocyte hypertrophy through mammalian target of rapamycin complex 1 pathway. J Endocrinol (2009) 202:299–307. doi:10.1677/JOE-09-0044

68. Estrada M, Espinosa A, Muller M, Jaimovich E. Testosterone stimulates intracellular calcium release and mitogen-activated protein kinases via a G protein-coupled receptor in skeletal muscle cells. Endocrinology (2003) 144:3586–97. doi:10.1210/en.2002-0164

69. Estrada M, Uhlen P, Ehrlich BE. Ca2+ oscillations induced by testosterone enhance neurite outgrowth. J Cell Sci (2006) 119:733–43. doi:10.1242/jcs.02775

70. Estrada M, Espinosa A, Gibson CJ, Uhlen P, Jaimovich E. Capacitative calcium entry in testosterone-induced intracellular calcium oscillations in myotubes. J Endocrinol (2005) 184:371–9. doi:10.1677/joe.1.05921

71. Estrada M, Liberona JL, Miranda M, Jaimovich E. Aldosterone- and testosterone-mediated intracellular calcium response in skeletal muscle cell cultures. Am J Physiol Endocrinol Metab (2000) 279:E132–9. Available from: http://ajpendo.physiology.org/content/279/1/E132.long

72. Shen J, Yu WM, Brotto M, Scherman JA, Guo C, Stoddard C, et al. Deficiency of MIP/MTMR14 phosphatase induces a muscle disorder by disrupting Ca(2+) homeostasis. Nat Cell Biol (2009) 11:769–76. doi:10.1038/ncb1884

73. Weisleder N, Brotto M, Komazaki S, Pan Z, Zhao X, Nosek T, et al. Muscle aging is associated with compromised Ca2+ spark signaling and segregated intracellular Ca2+ release. J Cell Biol (2006) 174:639–45. doi:10.1083/jcb.200604166

74. Chen JC, Goldhamer DJ. Skeletal muscle stem cells. Reprod Biol Endocrinol (2003) 1:101. doi:10.1186/1477-7827-1-29

75. Grounds MD, White JD, Rosenthal N, Bogoyevitch MA. The role of stem cells in skeletal and cardiac muscle repair. J Histochem Cytochem (2002) 50:589–610. doi:10.1177/002215540205000501

76. Charge SB, Rudnicki MA. Cellular and molecular regulation of muscle regeneration. Physiol Rev (2004) 84:209–38. doi:10.1152/physrev.00019.2003

77. Li Z, Mericskay M, Agbulut O, Butler-Browne G, Carlsson L, Thornell LE, et al. Desmin is essential for the tensile strength and integrity of myofibrils but not for myogenic commitment, differentiation, and fusion of skeletal muscle. J Cell Biol (1997) 139:129–44. doi:10.1083/jcb.139.1.129

78. Barberi L, Scicchitano BM, De Rossi M, Bigot A, Duguez S, Wielgosik A, et al. Age-dependent alteration in muscle regeneration: the critical role of tissue niche. Biogerontology (2013) 14:273–92. doi:10.1007/s10522-013-9429-4

79. Alway SE, Myers MJ, Mohamed JS. Regulation of satellite cell function in sarcopenia. Front Aging Neurosci (2014) 6:246. doi:10.3389/fnagi.2014.00246

80. Fukada S, Higuchi S, Segawa M, Koda K, Yamamoto Y, Tsujikawa K, et al. Purification and cell-surface marker characterization of quiescent satellite cells from murine skeletal muscle by a novel monoclonal antibody. Exp Cell Res (2004) 296:245–55. doi:10.1016/j.yexcr.2004.02.018

81. Hirai H, Verma M, Watanabe S, Tastad C, Asakura Y, Asakura A. MyoD regulates apoptosis of myoblasts through microRNA-mediated down-regulation of Pax3. J Cell Biol (2010) 191:347–65. doi:10.1083/jcb.201006025

82. Sinha-Hikim I, Roth SM, Lee MI, Bhasin S. Testosterone-induced muscle hypertrophy is associated with an increase in satellite cell number in healthy, young men. Am J Physiol Endocrinol Metab (2003) 285:E197–205. doi:10.1152/ajpendo.00370.2002

83. Wittert GA, Chapman IM, Haren MT, Mackintosh S, Coates P, Morley JE. Oral testosterone supplementation increases muscle and decreases fat mass in healthy elderly males with low-normal gonadal status. J Gerontol A Biol Sci Med Sci (2003) 58:618–25. doi:10.1093/gerona/58.7.M618

84. Yialamas MA, Hayes FJ. Androgens and the ageing male and female. Best Pract Res Clin Endocrinol Metab (2003) 17:223–36. doi:10.1016/S1521-690X(03)00018-6

85. Bross R, Javanbakht M, Bhasin S. Anabolic interventions for aging-associated sarcopenia. J Clin Endocrinol Metab (1999) 84:3420–30. doi:10.1210/jcem.84.10.6055

86. Cornelison DD, Wold BJ. Single-cell analysis of regulatory gene expression in quiescent and activated mouse skeletal muscle satellite cells. Dev Biol (1997) 191:270–83. doi:10.1006/dbio.1997.8721

87. Montero-Fernandez N, Serra-Rexach JA. Role of exercise on sarcopenia in the elderly. Eur J Phys Rehabil Med (2013) 49:131–43. Available from: http://www.minervamedica.it/en/journals/europa-medicophysica/article.php?cod=R33Y2013N01A0131

88. Brodsky IG, Balagopal P, Nair KS. Effects of testosterone replacement on muscle mass and muscle protein synthesis in hypogonadal men – a clinical research center study. J Clin Endocrinol Metab (1996) 81:3469–75. doi:10.1210/jc.81.10.3469

89. Bhasin S, Woodhouse L, Casaburi R, Singh AB, Mac RP, Lee M, et al. Older men are as responsive as young men to the anabolic effects of graded doses of testosterone on the skeletal muscle. J Clin Endocrinol Metab (2005) 90:678–88. doi:10.1210/jc.2004-1184

90. Morales A, Black A, Emerson L, Barkin J, Kuzmarov I, Day A. Androgens and sexual function: a placebo-controlled, randomized, double-blind study of testosterone vs. dehydroepiandrosterone in men with sexual dysfunction and androgen deficiency. Aging Male (2009) 12:104–12. doi:10.3109/13685530903294388

91. Snyder PJ, Peachey H, Berlin JA, Rader D, Usher D, Loh L, et al. Effect of transdermal testosterone treatment on serum lipid and apolipoprotein levels in men more than 65 years of age. Am J Med (2001) 111:255–60. doi:10.1016/S0002-9343(01)00813-0

92. Marks LS, Mazer NA, Mostaghel E, Hess DL, Dorey FJ, Epstein JI, et al. Effect of testosterone replacement therapy on prostate tissue in men with late-onset hypogonadism: a randomized controlled trial. JAMA (2006) 296:2351–61. doi:10.1001/jama.296.19.2351

93. Marks LS, Hess DL, Dorey FJ, Macairan ML. Prostatic tissue testosterone and dihydrotestosterone in African-American and white men. Urology (2006) 68:337–41. doi:10.1016/j.urology.2006.03.013

94. Percheron G, Hogrel JY, Denot-Ledunois S, Fayet G, Forette F, Baulieu EE, et al. Effect of 1-year oral administration of dehydroepiandrosterone to 60- to 80-year-old individuals on muscle function and cross-sectional area: a double-blind placebo-controlled trial. Arch Intern Med (2003) 163:720–7. doi:10.1001/archinte.163.6.720

95. Orr R, Fiatarone Singh M. The anabolic androgenic steroid oxandrolone in the treatment of wasting and catabolic disorders: review of efficacy and safety. Drugs (2004) 64:725–50. doi:10.2165/00003495-200464070-00004

96. Grunfeld C, Kotler DP, Dobs A, Glesby M, Bhasin S. Oxandrolone in the treatment of HIV-associated weight loss in men: a randomized, double-blind, placebo-controlled study. J Acquir Immune Defic Syndr (2006) 41:304–14. doi:10.1097/01. qai.0000197546.56131.40

97. Gao W, Reiser PJ, Coss CC, Phelps MA, Kearbey JD, Miller DD, et al. Selective androgen receptor modulator treatment improves muscle strength and body composition and prevents bone loss in orchidectomized rats. Endocrinology (2005) 146:4887–97. doi:10.1210/en.2005-0572

98. Papanicolaou DA, Ather SN, Zhu H, Zhou Y, Lutkiewicz J, Scott BB, et al. A phase IIA randomized, placebo-controlled clinical trial to study the efficacy and safety of the selective androgen receptor modulator (SARM), MK-0773 in female participants with sarcopenia. J Nutr Health Aging (2013) 17:533–43. doi:10.1007/s12603-013-0335-x

99. Dalton JT, Barnette KG, Bohl CE, Hancock ML, Rodriguez D, Dodson ST, et al. The selective androgen receptor modulator GTx-024 (enobosarm) improves lean body mass and physical function in healthy elderly men and postmenopausal women: results of a double-blind, placebo-controlled phase II trial. J Cachexia Sarcopenia Muscle (2011) 2:153–61. doi:10.1007/s13539-011-0034-6

100. Sacco A, Doyonnas R, Kraft P, Vitorovic S, Blau HM. Self-renewal and expansion of single transplanted muscle stem cells. Nature (2008) 456:502–6. doi:10.1038/nature07384

101. Singh R, Artaza JN, Taylor WE, Gonzalez-Cadavid NF, Bhasin S. Androgens stimulate myogenic differentiation and inhibit adipogenesis in C3H 10T1/2 pluripotent cells through an androgen receptor-mediated pathway. Endocrinology (2003) 144:5081–8. doi:10.1210/en.2003-0741

102. Singh R, Bhasin S, Braga M, Artaza JN, Pervin S, Taylor WE, et al. Regulation of myogenic differentiation by androgens: cross talk between androgen receptor/beta-catenin and follistatin/transforming growth factor-beta signaling pathways. Endocrinology (2009) 150:1259–68. doi:10.1210/en.2008-0858

103. Singh R, Artaza JN, Taylor WE, Braga M, Yuan X, Gonzalez-Cadavid NF, et al. Testosterone inhibits adipogenic differentiation in 3T3-L1 cells: nuclear translocation of androgen receptor complex with beta-catenin and T-cell factor 4 may bypass canonical Wnt signaling to down-regulate adipogenic transcription factors. Endocrinology (2006) 147:141–54. doi:10.1210/en.2004-1649

Author Notes

CHAPTER 2

Acknowledgments

MFJV is a professor of geriatric medicine and a paid speaker and consultant to Abbott Nutrition. C. J. Alish, A. C. Sauer, and R. A. Hegazi are employees of Abbott Nutrition, Abbott Laboratories. The material presented in this paper is based on published clinical evidence and is not affected by any financial relationship.

CHAPTER 3

Conflict of Interest

The authors declare that there is no conflict of interests regarding the publication of this paper.

CHAPTER 4

Competing Interests

The authors' declare that they have no competing interests.

Author contributions

RHC conceived the study, and participated in its design and coordination, performed the statistical analysis and drafted the original manuscript. SM and SS participated in the design and coordination of the study. ND participated in the design and coordination of the study, and served as the study physician. RW provided key insight into the conception of the study and edited the final draft of the manuscript. All authors read and approved the final manuscript.

Acknowledgments

Supported by NIH SBIR grant and NIH "GO" grant (list numbers of grants). We extend our gratitude for the expert technical assistance of Cosby Jolley and Rick Williams. We also extend our sincere appreciation to our volunteers for their effort towards the completion of this demanding study.

Disclosure statement

We would like to extend our sincere appreciation to our volunteers for their effort towards the completion of this study. Drs. Coker, Deutz, and Wolfe, and Mr. Schultzer were compensated by Healthspan, LLC as consultants through funds from the NIH SBIR grant for this study.

CHAPTER 5

Conflict of Interest

The authors declare no conflict of interest.

CHAPTER 6

Competing Interests

YY, TACV, NAB, LB, MAT, and SMP declare that they have no competing interests.

Authors Contributions

YY and SMP designed the research; YY, TACV, NAB, MAT and SMP conducted the research; YY and SMP analyzed the data; YY, TACV, LB, and SMP wrote and edited the manuscript; SMP had primary responsibility for the final content. All authors read and approved the final content.

Acknowledgments

We are grateful to Todd Prior and Tracy Rerecich for their technical assistance during data collection.

Funding

This work was funded by a research award from the US Dairy Research Institute to SMP, Grants from the Canadian Natural Science and Engineering Research Council (NSERC) to SMP and a graduate scholarship to TACV, and The Canadian Institutes for Health Research (CIHR) to SMP. YY, TACV, NAB, MAT, and SMP have no conflicts of interest, financial or otherwise, to declare.

CHAPTER 7

Acknowledgments
The authors are supported by the Claude D. Pepper Center for Older Americans in Little Rock, AR. Baum is supported by a Pepper Center Pilot Study Award P30 AG028718.

Author Contributions
The authors wrote and reviewed the material together.

Conflict of Interest
The authors declare no conflict of interest.

CHAPTER 8

Conflict of Interest
No conflict of interests, financial, or otherwise are declared by the authors.

Acknowledgments
The authors wish to thank Professor Timothy Bloom, Centro Linguistico di Ateneo of the University of Urbino Carlo Bo, for a critical reading of the paper. This research was supported by the Italian Ministry of Health, "Ricerca finalizzata 2009" (Grant no. RF-2009-1532789).

CHAPTER 9

Acknowledgments
This work was supported by a research Grant-in-Aid for Scientific Research C (no. 23500778) from the Ministry of Education, Science, Culture, Sports, Science, and Technology of Japan.

CHAPTER 12

Conflict of Interest

The authors declare that the research was conducted in the absence of any commercial or financial relationships that could be construed as a potential conflict of interest.

Acknowledgments

This work was supported by the Fondo Nacional de Ciencia y Tecnología (FONDECYT) grant 1120259 (to Manuel Estrada). Javier Duran thanks Comisión Nacional de Ciencia y Tecnología (CONICYT) for a student fellowship.

Index

β-carotene, 9, 225

A

acceptable macronutrient distribution range (AMDR), 129–130
activities of daily living (ADL), 48, 152, 166, 227
adenosine diphosphate (ADP), 46, 54, 91, 95, 97
adenosine triphosphate (ATP), 45–47, 54, 60, 91, 95, 97, 148, 154, 157–158, 161, 196, 249
adequate intake (AI), 46, 56, 190
adipose, 69–71, 73, 75–77, 79–85, 197, 207, 218, 241
age, xvii–xviii, xxii–xxiv, 3–12, 14, 16, 20, 22–28, 30–32, 34–36, 42, 47, 49, 51–53, 56–57, 60–61, 63–64, 69–71, 75, 86–87, 89, 97–98, 107, 116, 119, 122, 127–130, 137–139, 145, 148–161, 164–165, 168, 171–173, 175–178, 180–181, 186–187, 197–200, 203, 205–206, 218, 221–222, 224, 226–228, 231–232, 236, 239–244, 249, 252–257, 261–267
aging, xviii, xxi–xxiv, 3, 5, 14–16, 19–20, 22, 24–26, 29–30, 32, 47, 49, 51–52, 56–58, 61–64, 69, 84–85, 87–91, 96–98, 100, 104, 122–124, 127–128, 130–131, 138, 140, 142, 144–154, 156–161, 163, 165, 167–168, 170–180, 185–186, 188, 194, 196–201, 204–208, 211, 216,

218–219, 221–224, 226, 229–236, 238, 243–244, 247, 249, 253–254, 256–257, 259–263, 265–268
  cellular aging, 149, 158, 223
albumin, 22, 25, 122, 242
alcohol, 71
allergies, 71
amino acid, xx, xxii, 5–6, 8, 10, 15, 36–38, 42–46, 53–59, 69–71, 73, 75, 77, 79, 81–93, 95, 97–106, 110–111, 114, 117, 119, 121–125, 131–139, 141–146, 176, 187, 193, 198, 209–211, 213–214, 216–219, 227, 236
  branched chain amino acids (BCAAs), 37–38, 42, 44–46, 54, 56, 88
AMP-activated protein kinase (AMPK), 157, 160–161, 164–165, 177, 179, 247
anabolic resistance, xx, 36, 42, 53, 55–56, 88–89, 97–98, 100, 105, 121, 124, 210–211, 214, 216, 220
anabolic threshold, xxii, 56, 121, 209–217
androgen, xxiii, 155, 239–241, 243–247, 249–253, 255–261, 263–265, 267–268
angiotensin-converting enzyme (ACE) inhibitors, 37, 191, 193–194, 203–204
antioxidant, xviii, xx, xxiii, 5, 7–9, 12, 15, 37–38, 49–50, 55, 62, 151, 163, 167–168, 174, 177, 181, 216, 218–219, 221–238, 249–250, 265

Printed in the United States
by Baker & Taylor Publisher Services